THE CRUEL DECEPTION

The Use of Animals in Medical Research

Dr Robert Sharpe

D1627973

THORSONS PUBLISHING GROUP

First published 1988

© DR ROBERT SHARPE 1988

British Library Cataloguing in Publication Data

Sharpe, Robert
  The cruel deception : the use of animals
  in medical research
  1. Laboratory animals
  I. Title
  507'.24   QL55

  ISBN 0-7225-1593-6

Published by Thorsons Publishers Limited, Wellingborough, Northamptonshire, NN8 2RQ, England.

Printed and bound in Great Britain by Mackays of Chatham PLC, Kent

10 9 8 7 6 5

CONTENTS

Dedication

To all those whose only crime is to be born nonhuman

'I cannot think of a single major medical breakthrough that was produced as a result of an animal experiment. I wonder how many more million animals have to be sacrificed before we abandon the useless and barbaric practice of animal experimentation.'

(Dr Vernon Coleman,
Fellow of the Royal Society of Medicine)

ACKNOWLEDGEMENTS

This book was made possible through the support of the International Association Against Painful Experiments on Animals. Grateful thanks are also due to all our friends at the Bristol and West of England Antivivisection Society. I would also like to express my gratitude to Colin Smith and Brian Gunn, former colleagues at the NAVS, for their constant encouragement and support. Their knowledge, dedication and unique sense of humour have inspired and sustained me over the years. Above all, my thanks to Lorraine, who persuaded me to write the book in the first place and whose constant encouragement enabled it to be completed, sometime this century. All this has been achieved despite persistant interruptions from our feline friend, Uncus. I forgive you.

RS

Hazleton Laboratories, 1980

Paraquat is applied to the intact and abraded skin of New Zealand white rabbits to see how much it takes to kill them. The dying animals have difficulty breathing and some are prostrate on the cage floor, totally unable to move. One animal suffers a large anal haemorrhage and another bleeds heavily from the penis.

Oxford University, 1981

University scientist D. A. Winfield describes sight deprivation experiments in which ten-day old kittens have their eyes stitched closed. One of the animals dies after 6 months, never having seen daylight.

Wickham Research Laboratories, 1982

Conscious rabbits are restrained in stocks for eye irritancy tests with a British Petroleum machine lubricant. Within an hour their conjunctivae are crimson red, swollen and discharging.

Porton Down, 1983

Chemical Defence scientists publish experiments in which rhesus monkeys are anaesthetized and shot in the head with high velocity missiles. Survival times ranged from 2 to 169 minutes.

Cambridge, 1984

Scientists report how 28-day old mice are injected with herpes virus directly into the brain. One virus proves inevitably fatal within six to nine days, brain tissue being severely inflamed. Most of the dying animals also suffer eye infections and corneal ulcers.

Addenbrookes Hospital, 1985

Mongrel dogs undergo kidney transplant experiments but after 12 weeks less than half the animals are still alive.

FOREWORD

George Bernard Shaw said, 'If we attempt to contravert a vivisectionist by showing that the experiment that he has performed has not led to any useful result, you imply that if it led to a useful result, you would consider his experiment justified. Now I'm not prepared to concede that position.' I would agree to that – I know that so much terror and pain can never be warranted. But the vivisection industry is very powerful and *needs* to be challenged on its own ground.

I've never been inside an animal laboratory, but I remember the first time I saw some film footage taken inside one. After a few seconds of watching those fleeting visions of institutionalized hell on earth, I heard myself screaming. I knew for certain that what I was watching was wrong, could *never* be justified, but justify it people did. And what could I, a lay person, answer to their argument that the pain and anguish of all these animals has alleviated the pain and anguish of equal amounts of human beings? That argument has worked so well over the last hundred years that the vivisection industry has been allowed to grow, virtually unabated, into the massive, though hidden, institution it is today. That argument with its 'rational' justifications, has been hard to challenge with convictions that were rooted in an 'emotional' response to suffering. Many people see the world we humans inhabit as a purely *human* world – a world in which other species incorporated in it, are intrinsically of lesser value than human beings. Lacking the information to argue on their terms I have been thrown back on the simple conviction that there can be no more moral justification for slowly slicing up millions of laboratory animals than there can be for slicing up one domestic puppy. Antivivisectionists agree; vivisectionists do not.

Finally, with the arrival of Dr Sharpe's wonderful book, I am armed with evidence to argue what I have always suspected –

that it is *not* to animal experiments we owe the major medical discoveries that have so profoundly changed our lives. With detailed documentary and historical evidence (somehow Dr Sharpe manages to avoid ever being boring) the author carefully unravels the screen of myth and propaganda built around the vivisection establishment, at the heart of which lies the assumption that Animal Research is the Key Weapon in the Fight Against Disease. He not only exposes the falsehood of this, showing convincingly that it is *clinical* investigation together with intelligent application of observation that have provided most of the really important medical discoveries, but he makes abundantly clear how dangerous animal tests can be to humans. The disasters of Opren, Eraldin, or indeed something as simple as the tragically addictive effects of Valium on human beings, are sad examples of the ineffectiveness of animal testing. There are a great deal more.

Millions of animals die in anguish for reasons even less justified than the marketing of new drugs. They are sacrificed in tests on behalf of competing brands of floor cleaner or shampoo. They are deliberately put through agony by men studying aggression. They are starved to monitor the effects of hunger on sexual appetite. Huge resources are poured into these obscene studies, while outside in the real world more people are starving than ever before and civil violence is frighteningly on the increase.

Throughout history there has been dissent within the scientific community itself – it being as representative a community as any other – but more often than not those protestors lose their positions, as we read happened to Donald Barnes when he protested the futility of many of the gruesome experiments being carried out on animals in military laboratories. No doubt Dr Sharpe will also be attacked. I believe he will also be widely praised. There is so much in his book to make you think, reassess; it opens so many shut doors, that I would urge people of all spheres and interests to read it. At each turn it touches on something that must deeply affect someone. Mercifully, a book has been written from a scientific, rational point of view that might possibly serve to alleviate the mass of terrible and pointless suffering endured daily by laboratory animals.

JULIE CHRISTIE

INTRODUCTION

Scientists love animals

'Britain's 12,000 scientists who use animals in biological research would rather not . . . because they, too, love animals.'

(Research Defence Society[1])

Every year in Britain alone millions of animals suffer and die in laboratory experiments. They are burnt, scalded, poisoned and starved, given electric shocks and addicted to drugs; they are subjected to near freezing temperatures, reared in total darkness from birth and deliberately inflicted with diseases like arthritis, cancer, diabetes, oral infections, stomach ulcers, syphilis, herpes and 'AIDS'. Their eyes are surgically removed, their brains damaged and their bones broken. In military research they are gassed, poisoned with cyanide, shot with plastic bullets and deliberately wounded with high velocity missiles. There is no need to exaggerate – the scientists' own reports are enough.

For over a century the freedom to use living animals as tools of research has been protected by the Cruelty to Animals Act of 1876. The Act only dealt with experiments 'calculated to inflict pain' and during 1986 3,112,051 such experiments, mostly without anaesthetics, were reported to the Home Office.[2] There has never been a conviction for cruelty and the reason is simple. The Act is designed to protect the scientist and not the animal.

In 1986 the old legislation was replaced by the less emotive sounding Animals (Scientific Procedures) Act but the principle remains the same. Yet scientists have never been content with their legal protection and in 1908 established the Research Defence Society to promote actively the case for experiments on animals. Do not be fooled by the name – the Society does not defend research in general, only animal-based research. It appears convinced that animal experiments are vital to our health.

In 1937 the Honorary Treasurer of the RDS, Sir Leonard Rogers, concluded that antivivisectionists, in calling for an end to animal experiments, were . . . 'attempting the most wholesale cruelty to man and beast in the history of the world.'[3] Today's spokesmen are a little more restrained but the message is the same. W. D. Paton, a Vice President of the Society and a recent member of the Government's Advisory Committee on Animal Experiments, argues that, 'Animal experiment is essential to medical and veterinary research. Those who unnecessarily harass and restrict it now will carry a grave responsibility for the unnecessary ignorance and unnecessary human and animal suffering that will result in the future.'[4]

In 1985 the Honorary Secretary, Professor Tim Biscoe, whose special interest is the nervous system of mice, stated, 'Only by animal experiments have there been such huge advances in general medical and surgical care.'[5]

The Government's position was made crystal clear during recent Parliamentary debates on new legislation to replace the Cruelty to Animals Act. Amendments to ban the use of animals in several well-publicized areas of public concern, such as cosmetics tests, tobacco and alcohol research, warfare experiments and eye irritancy tests, were all rejected. In each case the Government strongly defended the use of animals. Indeed, it is often difficult to distinguish RDS statements from Government statements. Contrast the speech by Lord Glenarthur, Home Office Minister in the House of Lords, with the RDS statement quoted at the beginning of this chapter:

> '. . . I have been profoundly struck by the sense of responsibility and ethics displayed by all those who care for animals in science at every stage; those responsible for their husbandry, and those who undertake the detailed, painstaking and very advanced work on them. On visits I have made, their commitment not only to their research but to the well-being of the animals under research, stands out a mile. They are surely all lovers of animals.'[6]

But if vivisection is so harmless to the animals and so beneficial to the public why is the whole issue shrouded in secrecy, actively encouraged from the Home Office downwards?

The Government not only refuses to divulge the nature and scale of research carried out by individual laboratories but even the location of the 442 premises where animal experiments are performed. Information must be gleaned from the scientific literature, to which the public does not always have access. Even so, the British Government's own Littlewood Committee Report on Animal Experiments,[7] published in 1965, revealed that three out of every four experiments were not published at all '... because the results were inconclusive or the licensees judged them insufficiently important.' The same Report accepts there has been an '... appearance of secrecy about the practice of animal experimentation' and states that Home Office inspectors '... have tended to discourage laboratory authorities from inviting individuals or the Press to enter animal houses.' The desire for secrecy is further revealed by guidance notes obtainable from the Research Defence Society and '... compiled with the collaboration of the Home Office Cruelty to Animals Act Inspectorate':

> '... aim at a closed community in a self-contained unit with private lift or entrance(s) for staff, animal foodstuffs, etc.; not overlooked or, if so, fitted with opaque windows (not blinds) ... Premises selected and prepared for experimental animal usage should ideally be in a quiet place undisturbed by traffic and out of sight of the general public and afforded minimal publicity.'[19]

And just to make sure, the Government's own Advisory Committee on Animal Experiments is heavily overloaded with those who strongly support vivisection. Of concern too is Clause 24 of the new Animals (Scientific Procedures) Act, which prohibits the unlawful disclosure of confidential information. It is now even less likely that scientists will expose the suffering or futility behind their own or their colleagues' experiments.

Is it possible that powerful vested interests are at stake that must be protected? Suspicions increase when the claims for animal-based research are examined more closely.

In recent years animal experiments have been carried out on a huge scale – over 144 million in Britain alone since 1950, and if vivisection were making such an enormous contribution we could confidently expect a massive improvement in health. In fact, life-expectancy for the middle-aged has hardly changed since 1950;[8] the level of chronic sickness is extremely high and actually increasing;[9] more working days are being lost,[10] and hospital admissions are rising.[18] The number of prescriptions issued per person is increasing[11] whilst heart disease has reached epidemic proportions and cancer shows no signs of decline.[12]

The overwhelming majority (85 per cent) of animal experiments carried out over the past hundred years were performed since 1950, yet, by this time, the dramatic fall in death rate that commenced in the mid nineteenth century had largely levelled out.[13] In fact, over 92 per cent of the decline in deaths had been achieved by 1950. Furthermore, life-expectancy of American men ranked only nineteenth in a survey of 32 developed nations,[14] yet the United States uses more animals than anyone else – 65 to 100 million each year.

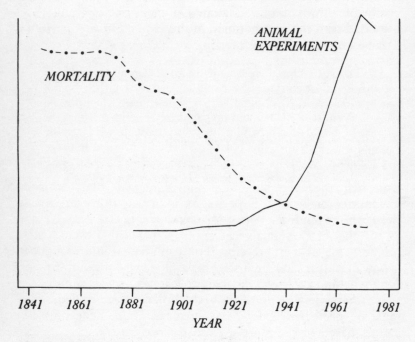

If animal tests are protecting us so well, why have drug side-effects reached epidemic proportions? A staggering 120,366 patients were discharged from or died in UK hospitals during 1977 after suffering the harmful effects of drug treatment.[15] Medical sources reveal that prescribed drugs cause between 3 per cent and 8 per cent of hospital admissions, whilst as many as 40 per cent of patients may experience side-effects in general practice.[16] Even so, no one knows the real level of drug-induced disease because side-effects are grossly under-reported.

Not only that, but animal-tested drugs are continually being withdrawn or severely restricted because of unexpected and often lethal effects. Recent examples include Opren, Zomax, Flosint, Osmosin, Zelmid, clioquinol, Ibufenac, chloramphenicol, Selacryn and Eraldin.[17]

In the past, people had to rely on bland assurances that animal experiments were strictly controlled, of enormous benefit, and, in any case, the scientists had the welfare of the animals at heart. But are the benefits really so great and can discoveries be made without the abuse of animals? In the chapters ahead the real factors that determine our health are revealed, together with the complex web of vested interests which conspire to keep Britain unhealthy. We explain how animal experiments are not only unnecessary but dangerously misleading. And we show how vivisection, far from protecting us against hazardous drugs, actually adds to the burden of disease.

1 RDS literature
2 'Statistics of experiments on living animals', 1986 Home Office (HMSO, 1987)
3 L. Rogers, *The Truth About Vivisection* (Churchill Livingstone, 1937)
4 W. Paton, *Man and Mouse* (Oxford University Press, 1984)
5 T. Biscoe, *RDS, Newsletter*, May, 1985
6 Lord Glenarthur, House of Lords, 28 November, 1985
7 Report of the Departmental Committee on Experiments on Animals, Home Office (HMSO, 1965)
8 *Social Trends*, no.15, 1985
9 *Social Trends*, no.14, 1984 and reference 8
10 Overall sickness absence from work has increased by 43 per cent from 1954/55 to 1978/79: N. Wells in G. Teeling Smith (Ed.), *Costs and Benefits of Pharmaceutical Research* (Office of Health Economics, 1987)
11 *SCRIP*, 2, 9 September, 1987
12 See Chapter 2
13 OPCS, Trends in Mortality 1951-75, Series DHI no.3 (HMSO, 1978)
14 On the State of the Public Health, 1979, DHSS (HMSO, 1980)
15 R. D. Mann, *Modern Drug Use—an enquiry on Historical Principles*, (MTP Press Ltd, 1984)
16 M. D. Rawlins, *BMJ*, 974-976, 21 March 1981
17 See Chapter 3
18 See references 52 and 81 in Chapter 2.
19 'Guidance notes on the law relating to experiments on animals' (Research Defence Society, 1974)

PART 1

Prevention is better than cure

CHAPTER 1

Out of the Dark Ages

'... it is a widely held fallacy that mortality from infectious diseases only commenced to fall with the advent of modern chemotherapeutic drugs.'

Ramsay & Emond, *Infectious Diseases* [1]

History reveals a constant ebb and flow in the pattern of infections and the emergence of new diseases. Plague was predominant in England for three centuries, then unaccountably disappeared. In Europe leprosy reached a peak between 1,000 and 1,400 AD then slowly declined, whilst smallpox held the stage for many years before gradually falling away. It is true that diseases often become less virulent and no longer a threat to life but, even so, at the beginning of the nineteenth century, diseases that had plagued the human race for centuries still took a massive toll of victims. Epidemics of typhoid, tuberculosis, whooping cough, scarlet fever, diphtheria and smallpox were a feared and common occurrence. Infant mortality was appalling and in London between 1790 and 1809 two out of every four children died before the age of five.[2] According to one estimate more than half the human race died before the age of ten.[3]

Action was desperately needed and it was Jeremy Bentham, a key figure in the new libertarian school of philosophy, who was to provide the spiritual inspiration for reform. Bentham's philosophy was to promote the greatest happiness of the greatest number and he was committed to reforms on a wide variety of issues. His enlightened attitude towards animal cruelty led to a forward-looking passage written in 1789, at a time when black slaves had been freed by the French but in the British dominions were still being treated in the way we now treat animals:

'The day *may* come when the rest of the animal creation may acquire those rights which never could have been withholden from them but by the hand of tyranny. The French have already discovered that the blackness of the skin is no reason why a human being should be abandoned without redress to the caprice of a tormentor. It may one day come to be recognized that the number of legs, the villosity of the skin, or the termination of the *os sacrum* are reasons equally insufficient for abandoning a sensitive being to the same fate. What else is it that should trace the insuperable line? Is it the faculty of reason, or perhaps the faculty of discourse? But a full-grown horse or dog is beyond comparison a more rational, as well as a more conversable animal, than an infant of a day or a week or even a month, old. But suppose they were otherwise, what would it avail? The question is not, Can they *reason?* nor Can they *talk?* but, Can they *suffer?*'[4]

Since health is an essential requirement for the people's happiness, Bentham set out to achieve the reforms so desperately needed. Amongst his disciples was Edwin Chadwick, a lawyer with an interest in social problems – just the man to overcome the active resistance and apathy, if Bentham's ideas were to be implemented. The passing of a Poor Law Act in 1834 provided Chadwick with his first real opportunity of applying pressure. A number of Commissioners were appointed under the Act whose duty it was to investigate the administration of the Poor Laws, and Chadwick was nominated Secretary to the Commission. Shortly afterwards they reported on the health of London's citizens and revealed a truly appalling state of affairs.[2] In one London parish of 77,000 people, 14,000 suffered every year from some sort of fever and 25 per cent of those who went down with fever, died of it!

The conditions in other parts of England were equally bad. Investigations revealed terrible housing conditions, inadequate water supplies, frequent pollution of what supplies there were, and considerable deficiencies in sewerage. Streets and cellars were full of human excrement and every kind of refuse, and pools of fetid water stood everywhere. In his report to the government, published in 1842, Chadwick stated:

'That the various forms of epidemic, endemic and other disease caused, or aggravated, or propagated chiefly among the labouring classes by atmospheric impurities produced by decomposing and vegetable substances, by damp and filth, and close and overcrowded dwellings prevail among the population in every part of the Kingdom, whether dwelling in separate houses, in rural villages, in small towns, in the larger towns – as they have been found to prevail in the lowest districts of the metropolis;

'That such disease, wherever its attacks are frequent, is always found in connection with the physical circumstances above specified ... and where the removal of the noxious agencies appears to be complete, such disease almost entirely disappears'[5]

Chadwick's recommendations included drainage, removal of refuse from houses, streets and roads, improvement of water supplies, a bold housing policy and improvements in the standard of living and in working conditions.[2] By tackling causes rather than effects, it would cost society less to prevent the diseases than to cope with its victims. The conclusions were so obvious that eventually Parliament was forced to take action. The resulting Public Health Act of 1848 therefore represents the true renaissance of public health in Great Britain.[2] Its basic idea was that many diseases could be avoided by continuous action directed towards the creation of a healthy environment and it proved the first of many subsequent legislative measures. Nevertheless, it was the 1875 Public Health Act that acted far more positively in promoting sanitary reform.

In America, Lemual Shattuck led the way and in 1850 published a survey of the sanitary conditions in Massachusetts. Shattuck found great inequalities in the life-span between one social class and another and concluded:

'We believe that the conditions of perfect health, either public or personal, are seldom or never attained though attainable – that the average length of human life may be very much extended, and its physical power greatly augmented – that in every year, within this Commonwealth, thousands of lives are lost which might have been saved – that tens of thousands of cases of sickness occur, which might have been prevented – that a vast amount of unnecessarily impaired health, and physical debility exists among those not actually confined by sickness – that these preventable evils require an enormous expenditure and loss of money, and impose upon the people unnumbered and unmeasurable calamities, pecuniary, social, physical, mental and moral, which might have been avoided – that means exist, within our reach, for their mitigation or removal – and that measures for prevention will effect infinitely more than remedies for the cure of disease.'[5]

Over the next hundred years Britain's death rate fell rapidly, due almost exclusively to the decline of the infections – mainly TB, bronchitis, pneumonia, influenza, whooping cough, measles, scarlet fever, diphtheria, smallpox, cholera, typhoid, diarrhoea and dysentery.[6] As we shall see, mortality for virtually all the infections was declining before, and in most cases long before, specific therapies became available. The impetus to better health from the mid nineteenth century onwards can therefore be directly traced to public health measures and social legislation that improved the living standards of working people. Higher wages and welfare benefits made it possible for the poor to eat properly and public health measures radically improved conditions in the densely-populated urban areas, particularly with the provision of clean water supplies, sanitation, sewerage and new housing. Susceptibility to the infections diminished rapidly as nutrition, housing, hygiene and general living conditions improved. In 1902 the Midwives Act was passed which improved the training and skills of midwives and from the turn of the century infant mortality fell rapidly, in line with the improvement in social conditions.[7]

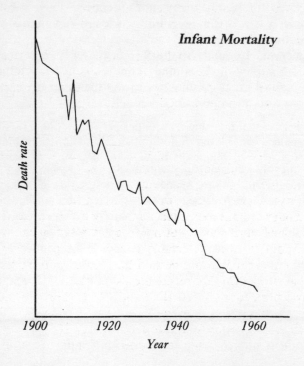

Infant Mortality

In fact infant mortality is now regarded as one of the best indications of a country's social conditions.[7] Infant mortality is lowest in countries with a very high standard of living such as Sweden and Norway with 7.8 and 8.6 deaths per 1,000 live births respectively, compared with poorer countries such as Yugoslavia, with a rate of 33.6.[8]

It may seem strange to us now but the enormous importance of public health had been recognized long ago in Roman times, and even before that.[2] Principal cities of antiquity prided themselves on their aqueducts, their sewers and their public baths. Burial grounds were situated outside built-up areas. The Romans paid great attention to the drainage, sanitation and water supplies of their cities and Rome possessed its subterranean drains as early as the sixth century BC. Its magnificent water supplies were carried to the city by 14 great aqueducts with strict precautions taken against pollution. Regulations enforced the cleanliness of its streets. But after the fall of the Roman empire preventive medicine virtually disappeared and remained in abeyance for the whole of the Middle Ages, and for a long time afterwards. Finally the eighteenth and early nineteenth centuries saw the beginnings of sanitation in towns and new cemeteries were once again sited away from built-up areas.

Just how much we owe social reformers like Chadwick can be seen by analysing the declining death rates for each major infectious disease. A hundred years ago tuberculosis was a major killer and in the early 1850s claimed nearly 3,000 victims per million of the population.[6] Malnutrition, overcrowded and ill-ventilated conditions and poor housing all enabled the disease to flourish but, as conditions improved, death rates declined rapidly and were considerably reduced by the time specific treatment became available.[6]

Streptomycin, introduced in 1947, together with related drugs is thought to have speeded up the already declining death rate in England and Wales, but not in the United States where therapy produced little detectable change.[9] According to one estimate the most that could be said for medical intervention is that antibacterial drugs reduced the number of TB deaths in Britain by 3 per cent between 1848 and 1971.[6] The general use of BCG vaccination began in 1954 but its value is highly questionable. In a major trial in India involving 260,000 people, not only did it prove totally ineffective but more cases of TB

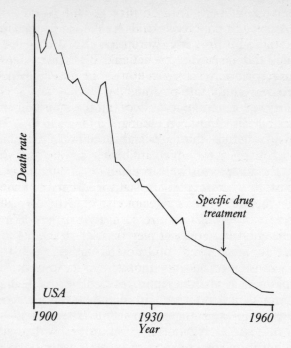

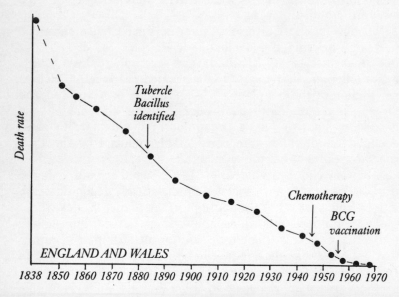

Tuberculosis in England, Wales and the USA

occurred in those vaccinated than in the placebo group![10] On the other hand, the Netherlands had the lowest death rate from respiratory TB for any European country in 1957-59 and 1967-69 despite having no national BCG programme.[6]

During the nineteenth century, pneumonia, bronchitis and influenza were all grouped together in national statistics. The introduction of antibiotics does not seem to have made an impact on the already declining death rate but this is hardly surprising because influenza and some cases of acute bronchitis are viral diseases for which antibiotics are ineffective. Although it is known from clinical experience that antibiotics can successfully treat pneumonia, statistics from 1900 both here and in the USA do not show a major change in the already declining death rate.[11] Since there are so many types of influenza virus, vaccines against one form may be useless against another and one study, involving 50,000 Post Office workers, showed that influenza vaccine had no impact on absenteeism.[12] But mass vaccination can sometimes prove dangerous and it was in 1976 that President Ford ordered the now infamous nationwide vaccination programme against swine flu. Eventually the project had to be abandoned because the vaccine was found to cause death and paralysis amongst the elderly.[10]

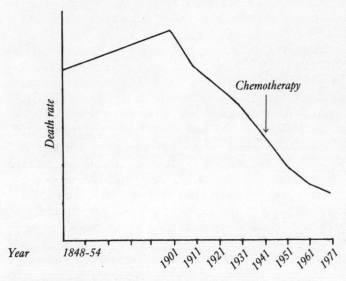

Bronchitis, pneumonia and influenza: death rates (standardized to 1901 population) for England and Wales

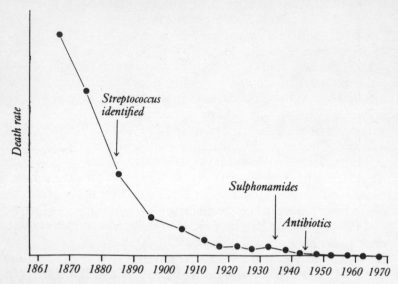

Scarlet fever: the mean annual death rate in children under 15 for England and Wales

Sadly pneumonia still claims many lives – 24,687 in England and Wales during 1984, and death rates for those aged 65 to 84 have risen sharply since the mid 1940s, despite the availability of antibiotics.[13]

Both scarlet fever and whooping cough have declined rapidly since the 1860s and 1870s and had fallen to comparatively low levels by the time antibiotics, and immunization against whooping cough, became available.[14] In the 1860s the death rate from whooping cough was about 1,372 per million children under 15. By 1901-10 it had fallen to 815, by 1921-30 to 405, and by 1940 to about 140 per million. In 1947-8 the rate had declined still further to 73 per million and by the time a nationwide vaccination programme was initiated in the late 1950s, the rate had fallen to around 5 per million children.[15] Since 1969 almost half the deaths have occurred in children under three months old[15] – before vaccination is commenced. In recent years the value and safety of the vaccine has been hotly disputed in the medical press and risks of brain damage between 1 in 750 and 1 in 100,000 children have been quoted.[16] In Glasgow Professor Gordon Stewart, a fierce critic of the vaccine, has found that a child's social class is three times more

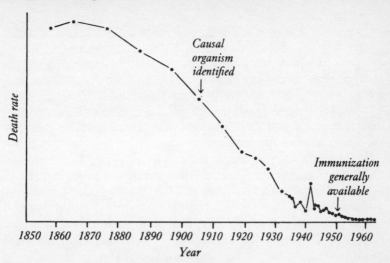

Whooping cough: death rates of children under 15 for England and Wales.

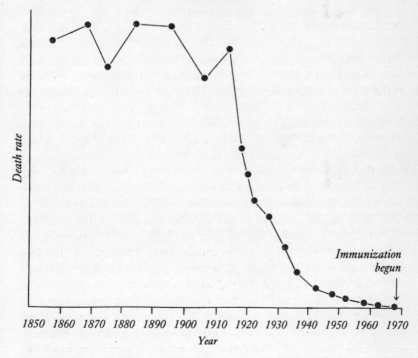

Measles: death rates of children under 15 for England and Wales

important than vaccination in influencing whooping cough outbreaks[17] – more evidence that poverty can cause disease. Other studies revealed that one third of whooping cough patients had previously been immunized![18] Routine use of the vaccine has been stopped in Sweden and West Germany without any rise in deaths or serious disease.[19]

Measles started to decline rapidly at the turn of the century and the death rate had reached very low levels by the time vaccination was introduced in 1968.[6]

In 1860 diphtheria accounted for well over 1,000 deaths per million children[20] but this had fallen sharply to an annual rate of around 400 between 1861 and 1870.[21] Although this fall was not associated with any specific therapy, later declines roughly coincided first with the introduction of horse antitoxin treatment (1894) and then by immunization (1940). Had mortality from other common childhood infections remained the same or increased during the same period, then it would be natural to assume that antitoxin and vaccination were mainly responsible for the fall in diphtheria deaths around 1900 and 1942. But deaths from whooping cough and measles did indeed decline over the same period without any treatment or immunization, suggesting other influences, such as an improved standard of living, may also have been at work with diphtheria. This is confirmed by figures from poorer countries where the death rate from diphtheria is 100 times higher.[22] And evidence taken from countries with a higher standard of living also shows that antitoxin and immunization could not have been solely responsible for the decline of diphtheria.

In his Presidential address to the British Association for the Advancement of Science, Professor Porter described how the value of antitoxin treatment has never been accepted generally[23] and perhaps this is not surprising because controlled clinical trials were never carried out.[24] As a result there are virtually no statistics proving that antitoxin actually works on human beings.[25] The apparent fall in the case-fatality rate (the number of deaths expressed as a fraction of the total number contracting the disease) may well have been caused by new diagnostic methods. Bacterial analysis, introduced at about the same time as antitoxin, meant that mild cases of the disease, previously classified as something else, were now included in statistics, which automatically lowered the overall case-fatality rate.[26] And despite the availability of antitoxin since the 1890s, several

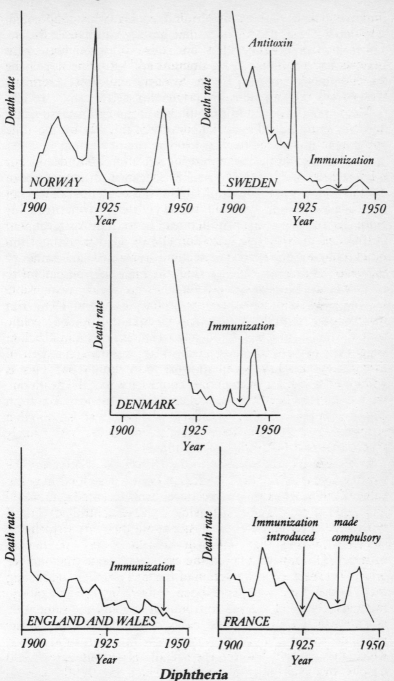

Diphtheria

countries have experienced an increased death rate in the early years of this century.[27] In Berlin during the 1920s a severe outbreak of diphtheria led to high case-fatality rates despite large doses of antitoxin being given at an early stage. Debating these findings at the Berlin Medical Society, Professor Friedberger argued that the apparently favourable results following the introduction of antitoxin in the 1890s were really due to a natural decline in the severity of the disease.[28]

Figures for America show that immunization against diphtheria did not produce any detectable change in the death rate, which had already steeply declined.[24] Furthermore diphtheria was gradually declining in Massachusetts, Michigan and New York from about 1880, well before the introduction of antitoxin, let alone vaccination. Studies have also shown that parallels between the number of children immunized and the decline in mortality do not hold for every region in the USA,[24] suggesting that other factors, such as improvements in living standards, are crucially important.

In countries with a high standard of living, such as Denmark, Sweden and Norway, deaths from diphtheria declined rapidly without any vaccination.[29] That is, until World War II when several countries in Western Europe had greatly increased rates. In Denmark, immunization did not begin until 1941, but in

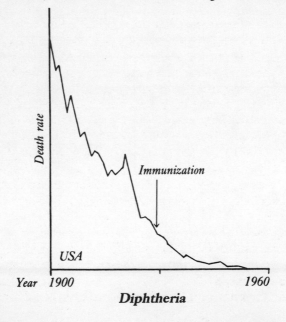

Diphtheria

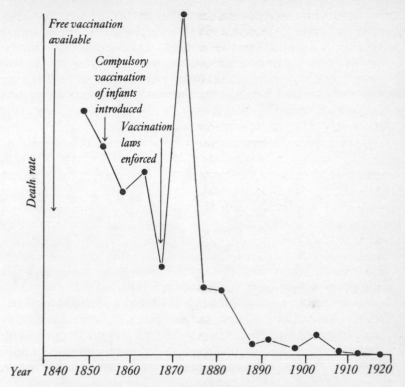

Free vaccination
available

Compulsory
vaccination
of infants
introduced

Vaccination
laws
enforced

Death rate

Year 1840 1850 1860 1870 1880 1890 1900 1910 1920

Smallpox: death rates for England and Wales

Copenhagen, despite 95% of children being inoculated, there was an astonishing increase from 41 cases in 1942 to 1,754 cases in 1944![30] In Norway the disease had rapidly declined and virtually disappeared by 1939 when only 18 cases per million were recorded.[31] It was only after the German occupation in World War II that immunization was introduced, coincidentally with an enormous rise in diphtheria. And, despite immunization, diphtheria had shown a remarkable rise in Germany both before and during the Second World War. Increased over-crowding, a general lowering of hygienic standards and a lack of resistance because of poor food supply seem largely responsible.[31]

Evidence like this, taken from the experience of other countries shows that antitoxin and immunization could not have been solely, or even mainly, responsible for the decline of diphtheria in Britain.

Although the contribution of smallpox to the overall decline in

Britain's death rate between the 1850s and 1970s was relatively small, this is the one major disease for which vaccination was available before 1900. The medical historian Creighton considered vaccination against the disease useless but this is not a generally accepted view. Nevertheless, a recent analysis of the decline of smallpox in London concluded that vaccination could never have been solely responsible.

> 'The history of smallpox in the later years of the nineteenth century does not support the contention that vaccination was fully or finally responsible for the eventual disappearance of the disease in Britain. It was in these years, in fact, that there was developed the system for control of the disease that became the basis of the successful modern campaign for its eradication.'
>
> *Medical History*, 1983[32]

The 'system' included the establishment of port sanitary authorities to avoid the disease being imported from abroad, isolation of patients and thorough cleansing of their homes. In fact the Royal Commission on Smallpox and Fever Hospitals traced the beginnings of the decline to the 1780s although Jenner's method of vaccination was not even published until 1798. By the time Jenner's vaccine was being introduced, between 1801 and 1810, the death rate had already fallen from 500 to 200 per 100,000 of the population.[32] Even then vaccine uptake was not great – hence the subsequent Acts of Parliament attempting to enforce the practice. Compulsory vaccination was introduced in 1852 but by then mortality had fallen to 40 per 100,000. Between 1871 and 1880, the period when compulsory vaccination was legally enforced, the death rate leapt from 28 to 46 per 100,000.[32] Worldwide, the elimination of smallpox has been attributed to isolation of contacts, education and mass vaccination.[33]

It is often thought that Edward Jenner developed the first protection against smallpox but inoculation against the disease had been practised in India since ancient times and in China since 1063.[2] In 1718 Lady Wortley Montague, wife of the British Ambassador to Turkey, introduced inoculation against smallpox into this country. Small amounts of material from the pustules of those suffering from a mild form of the disease were administered nasally or inoculated into those seeking protection, immunity being conferred against dangerous attacks. The method was sometimes hazardous although risks could be reduced by making the fluid less virulent.

Jenner's subsequent 'discovery' of vaccination was really based on the chance observation that milkmaids, infected with cowpox from lesions on the udders of cows, were protected

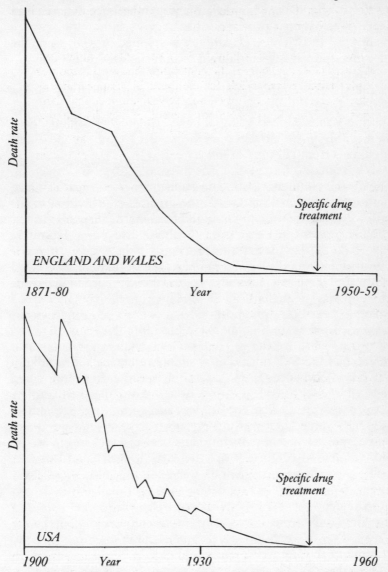

Typhoid Fever in England and Wales and the USA

against subsequent attacks of smallpox.[34] The naturally arising cowpox was similar to smallpox and in 1796, he tested the theory by inoculating material from cowpox lesions into a boy, John Phipps, who was subsequently found to be immune on deliberate inoculation with smallpox. Jenner's results were published in 1798 but didn't gain immediate acceptance because his claim that vaccination provided prolonged immunity proved incorrect.[1] In Britain vaccination was stopped in 1940 when the considerable risks were thought to outweigh the benefits.

Attempts to develop protection against smallpox have travelled full circle. Centuries ago, material from fellow sufferers was used whilst in more modern times smallpox vaccine was produced using animals. But recent findings have shown this to be unnecessary because vaccine can be safely produced from cultures of *human* cells.[35]

Whilst the decline of airborne infections like TB, bronchitis, pneumonia, influenza, whooping cough, measles, scarlet fever, diphtheria and smallpox, made the greatest contribution to the fall in deaths, the water-, milk- and food-borne diseases such as cholera, dysentery, diarrhoea, typhoid and non-respiratory tuberculosis, also played a major part. Cholera was introduced into Britain from the continent of Europe during the nineteenth century. The last epidemic was in 1865 and from that time cholera was negligible.[6] The decline of cholera is a good example of how practical advances are not necessarily dependent on scientific understanding. In 1854 John Snow, an early English anaesthetist, cut short a cholera epidemic that threatened the Soho district, by simply removing the handle of the Broad Street pump.[23] Careful observation had shown him that all the victims had drunk water from the well and he therefore deduced that cholera was spread by water contaminated with some kind of living organism.

The general decline of the diarrhoeal diseases began in the late nineteenth century. From the early 1900s, a rapid decline set in and by the time rehydration therapy was introduced in the 1930s to prevent dehydration, 95 per cent of the fall had already been achieved.[6] Non-respiratory TB, which can arise by drinking milk from cows with bovine tuberculosis, had also declined to a very low level by the time the first specific treatment, streptomycin, was introduced.[6]

The decline of enteric fevers – typhoid and paratyphoid, a major killer in the nineteenth century – was also rapid and began before the turn of the century. Specific treatment was not available until about 1950 but by then mortality had almost been eliminated both here and in the United States.[36]

Our analysis has naturally concentrated on the infectious diseases mainly associated with the fall in deaths and we have therefore excluded poliomyelitis. In 1947 when the highest ever death rate was recorded in England and Wales, there were 33 deaths per million children under 15, compared with 99 for whooping cough and 69 for measles.[14] In 1871-80 when the last two diseases were near their peak, 1415 and 1038 deaths per million children were recorded. This is not to minimize the importance of polio to those who contracted the disease but to show that its contribution to the overall decline in deaths has been small compared with other infections. In fact, deaths and the number of cases were both declining from the high point in 1947, so death rates were already low[37] when the Salk vaccine was introduced in 1956. A similar picture emerged in the United States[24] so it is difficult to assess the overall contribution of immunization in these countries. To what extent the epidemic of polio resulted from vaccination against other diseases is unknown but a report published in 1956 by Britain's Medical Research Council[38] revealed that 13 per cent of paralytic cases in children aged six months to two years resulted from vaccination against diphtheria and whooping cough! The report admitted this could well be an underestimate.

Later on, in 1962, Sabin introduced a new vaccine based on live but attenuated virus strains. This was considered more effective than the Salk vaccine but by then the disease had already fallen to very low levels. As described in Chapter 6, polio vaccine can be produced more safely from human cells in culture, so avoiding the dangers caused by vaccines made from animal tissues.

The evidence clearly shows that deaths from virtually all the infectious diseases were declining before, and in most cases long before, specific therapies became available. For the commonest infections of childhood – scarlet fever, measles, whooping cough and diphtheria – around 90 per cent of the total decline in mortality between 1860 and 1965 had already occurred *before* the introduction of antibiotics and widespread immunization against diphtheria. According to a classic analysis[39] by Professor Thomas McKcown of Birmingham University, the decline in mortality for the second half of the nineteenth century was due wholly to a reduction in deaths from infectious diseases, the

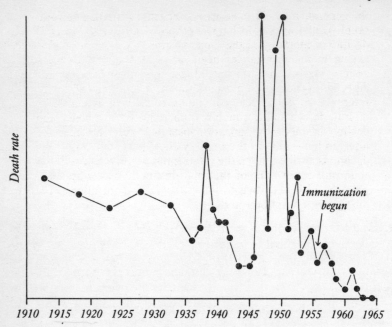

Poliomyelitis: the mean annual death rate for children under 15 for England and Wales

main reasons being a rising standard of living of which the most important feature was a better diet, improvements in hygiene, and a favourable trend in the relationship between some microbes and the human host. McKeown concludes that therapy made no contribution and the effect of immunization was restricted to smallpox, a disease that accounted for only about one twentieth of the reduction of the death rate. For the twentieth century McKeown lists improved nutrition and better hygiene as by far the most important influences, with therapy and immunization playing a relatively minor role. In the United States where Shattuck had performed the same vital role as Chadwick in Britain, researchers John and Sonja McKinlay found that for ten common infectious diseases (TB, scarlet fever, influenza, pneumonia, diphtheria, whooping cough, measles, smallpox, typhoid and poliomyelitis), medical measures only accounted for between 1 and 3.5 per cent of the decline in mortality since 1900.

'In general, medical measures (both chemotherapeutic and prophylactic) appear to have contributed little to the overall decline in mortality in the United States since about 1900 – having in many instances been introduced several decades after a marked decline had already set in and having no detectable influence in most instances. More specifically, with reference to those five conditions (influenza, pneumonia, diphtheria, whooping cough and poliomyelitis) for which the decline in mortality appears substantial after the point of intervention – and on the unlikely assumption that all of this decline is attributable to the intervention – it is estimated that at most 3.5 per cent of the total decline in mortality since 1900 could be ascribed to medical measures introduced for the diseases described here.'[9]

Even traditional medical sources such as the *Lancet* acknowledge that '. . . public health legislation and related measures have probably done more than all the advances of scientific medicine to promote the well-being of the community in Britain and in most other countries.'[40].

But if affluent Western societies have largely eradicated the infectious epidemics, life is not so good in poorer countries. People in the Third World now suffer from the same communicable diseases that were widespread in developed nations during the nineteenth century. Many illnesses are transmitted by food and water contaminated by disease organisms from human and animal excreta. They include diarrhoeal disease, amoebic and bacterial dysentery, typhoid, cholera, polio and infectious hepatitis.[41] Fewer than one in five people in the Third World can obtain clean water. Lack of a clean water supply, and the breeding of mosquitoes and flies in stagnant water, have been connected with 80 per cent of disease in the world.[33] Indeed the World Health Organisation estimate that 25 million people die every year because they do not have clean water and sanitation. Poverty is another major cause of ill health leading to malnutrition and a lowered resistance to infection: the death rates from whooping cough and measles are 300 and 55 times higher respectively in poorer countries.[22] The Third World poor almost always live in overcrowded conditions which accelerates the spread of disease. Whilst modern drugs can tackle many of these infections, they are powerless to break the cycle of disease if the environment remains unhealthy. As Oxfam point out, disease that is rooted in poverty can only be prevented by an onslaught on poverty and inequality.[41] In the words of the Tanzanian Food and Nutrition Council,[42] a '. . . society that is perpetuating malnutrition cannot be treated with medicine. It has to develop and be restructured in such a way that all its members are ascertained all their basic human needs.' So the prescription for better health in Third World countries is the same as that which worked so effectively in developed nations like the UK: improve nutrition, hygiene and sanitation and living and working conditions. Even tropical diseases like malaria can be effectively controlled through public health measures, that is by draining swamps or treating water so mosquitoes cannot breed.[33]

The evidence shows that society's control of infectious disease rests primarily on efficient public health services and a good standard of living and the dramatic increase in life-expectancy since the early 1800s can be directly traced to these sources. Medical measures clearly played only a relatively small part and later on, in Chapter 5, we will see how little even these owed to experiments on animals. None of this is an argument against *properly conducted* medical research, but it does show that the major influences on our health are outside the scope of laboratory experiments. As medical historian Brian Inglis concludes:

'The chief credit for the conquest of the destructive epidemics ... ought to have been given to the social reformers who had campaigned for purer water, better sewage disposal and improved living standards. It had been their efforts, rather than the achievement of the medical scientists, which had been chiefly responsible for the reduction in mortality from infectious diseases.'[43]

1 A. M. Ramsay and R. T. Emond, *Infectious Diseases* (Heinemann, 1967)
2 R. Sand, *The Advance to Social Medicine* (Staples Press, 1952)
3 Professor R. Watt in reference 2
4 Reproduced in *Animal Liberation*, P. Singer (Thorsons Publishing Group, 1983)
5 Reproduced in reference 2
6 T. McKeown, *The Role of Medicine* (Blackwell 1979)
7 F. Grundy, *Preventive Medicine & Public Health* (H. K. Lewis, 1964)
8 On the state of the Public Health, 1979 (DHSS, 1980)
9 J. B. McKinlay & S. McKinlay, *Health & Society*, 405-428, 1977 (Millbank Memorial Fund)
10 M. Weitz, *Health Shock* (Hamlyn, 1982)
11 In England and Wales, reference 6; in the United States, reference 9

12 R. Smith, *Lancet*, 330, 10 August, 1974
13 See reference 82 in Chapter 2
14 T. McKeown and C. R. Lowe, *An Introduction to Social Medicine* (Blackwell Scientific Publications, 1976)
15 *BMJ*, 1208, April 9, 1983
16 For instance, G. T. Stewart, *BMJ*, 1263, 24 April, 1982, together with reference 19.
17 W. R. Bassili and G. T. Stewart, *Lancet*, 471-473, 28 February, 1976
18 T. T. Salmi, et al, *Lancet*, 811-812, 25 October, 1975
19 *Times*, 8 September, 1982
20 W. H. Parry, *Communicable Diseases* (Hodder & Stoughton, 1979)
21 MRC Special Report Series, no. 247, 1943
22 *Lancet*, 632, 14 September, 1974
23 Presidential address by R. R. Porter at the Swansea Meeting of the British Association for the Advancement of Science, 3 September, 1971
24 H. F. Dowling, *Fighting Infection* (Harvard University Press, 1977)
25 A. B. Christie, *Infectious Diseases* (Churchill Livingstone, 1980)
26 C.S. Singer and E. A. Underwood, *A Short History of Medicine* (Clarendon Press, 1962)
27 For example, H. J. Parish's *Victory with Vaccines* (Churchill Livingstone, 1968) describes how antitoxin was widely used in Germany and France by 1895-1900. Yet in the early years of the twentieth century death rates for diphtheria showed a huge rise. See also graph for Sweden.
28 *Lancet*, 598, 14 August, 1931
29 Diphtheria graphs plotted using statistics from *Epidemiological & Vital Statistics Report*, 92-111, Volume 4, 1951 (WHO). According to Sweden's National Central Bureau of Statistics, vaccination was '. . .introduced in 1939 but not used to a greater extent until 1943' (letter from G. Karlström, 12 August, 1986.) According to the *BMJ*, 614, 3 November, 1945, immunization had not been carried out in Norway before World War II because it '. . . had not been considered necessary'
30 *Lancet*, 915, 20 December, 1947
31 *Lancet*, 628, 11 November, 1944
32 A. Hardey, *Medical History*, 111-128, volume 27, 1983
33 The Wellcome Museum of the History of Medicine (Science Museum, London, November 1986)
34 See Chapter 5
35 L. Hayflick, *Laboratory Practice*, 58-62, volume 19, 1970
36 Typhoid decline in England and Wales: graph plotted using data from *Registrar General's Statistical Review*, 1970 (HMSO, 1972); for United States, see reference 24
37 For declining death rates from polio see graph (reference 14). Reference 1 indicates that the total number of cases had fallen from around 10,000 in 1950 to just over 3,000 by 1956
38 *Lancet*, 1223-1231, 15 December, 1956
39 Conclusion reproduced in reference 9. See also references 6 and 14
40 *Lancet*, 354-355, 12 August, 1978
41 D. Melrose, *Bitter Pills* (Oxfam, 1982)
42 Reproduced in reference 41
43 B. Inglis, *Diseases of Civilization* (Paladin Books, 1983)

CHAPTER 2

The diseases of civilization

'Today's main killing diseases are due to the way we live . . . Logically the main thrust of medical research should be directed at the prevention of these common, lethal and disabling conditions.'

(Editorial, *British Medical Journal*[1])

The virtual disappearance of the infectious epidemics has made Britain a safer place to live. And we owe the improvement, not to what happens when we are ill, but because we do not so often become ill; and we remain well, not because of specific measures such as vaccination, but because we enjoy a higher standard of nutrition and live in a healthier environment.[2] But if vivisection had little impact in the past, what of our present burden of disease? Can animal experiments expect to achieve major advances in the future?

Heart disease is Britain's number one killer, claiming over 150,000 lives every year.[3] More than 1.5 million middle-aged men (one in three) suffer from heart disease and over 56 million working days are lost every year as a result.[4] Before 1925 coronary heart disease was uncommon but then increased steadily[5]: between 1950 and 1974 deaths more than doubled in England and Wales.[6]

The way in which coronary disease kills its victims is well understood. Autopsy specimens show how the heart's vital arteries become 'furred up' with fatty deposits called atheroma and when this happens the heart is starved of oxygen.[7] This leads to angina but if a blood clot also forms in the blocked artery, the heart muscle receives no oxygen whatsoever and a heart attack results.

About 50 per cent of all fatal heart attack victims die within half an hour, often before medical help arrives.[8] Even so the best that can be said for intensive coronary care units is that they have reduced overall deaths by just 4 per cent,[6] although in one study they proved no better, and in some cases worse, than simply sending the patient home to be treated by the family doctor.[9]

Symptoms can sometimes be alleviated by coronary bypass surgery in which a healthy vein is taken from another part of the body and grafted into position to replace the blocked heart artery. To what extent this prolongs survival is difficult to assess but even on the most optimistic estimates, its impact on the overall death rate and level of sickness is thought to be small.[10] In a survey of 600 patients with chronic stable angina, surgery made no difference to survival but did reduce the severity of the symptoms,[11] whilst another study found no evidence of increased life-expectancy except in patients with severe obstruction of the left main heart artery.[12]

It is also difficult to see how heart transplants could possibly cope with Britain's epidemic of heart disease. In a letter to *The Times*, Dr Donald Gould expressed a view shared by many:

'Those of us who regard heart, liver and kidney transplants as bad medicine do so not on the grounds that surgeons are squandering scarce resources on a treatment that should still be regarded as experimental, but because of a recognition that spare part surgery of this kind can never make any significant impact upon the toll of premature deaths extorted by the disease concerned.'[13]

The sheer cost and volume of resources means that *at best* heart transplants can only prolong the life of a tiny handful of people. American figures released in 1977 showed that out of a total of 346 hearts inserted into 338 patients (some needed more than one transplant) only 77 were still alive at the time of the survey.[14] The longest survivor had held on for eight and a half years. In the UK, after a disastrous start, results improved and by September 1984, 221 patients had received transplants at Papworth and Harefield hospitals but the survival rate was still only 54 per cent after three years.[15] Behind the glamorous image and sensational headlines lies a more fundamental question. Are transplants, as the *Lancet*[16] put it '. . . a medical Concorde – technologically magnificent, socially erroneous?'

Death rates for coronary heart disease vary widely from country to country and show that drugs have little overall impact on reducing mortality. In Japan the death rate for men aged between 45 and 54 was only 28 per 100,000 of the population in 1978 whilst the corresponding rates for America, England and Wales, Scotland, and Finland were 272, 272, 353 and 392 respectively.[5] Both the UK and Japan have similar access to the wide range of available heart drugs and if therapy were having such a marked effect, death rates in this country would rapidly decline towards those found in Japan. In fact deaths from heart disease have remained practically constant in Britain since the early 1970s and the Japanese owe their low rates not to their genes but to their way of life because when they move to America, they quickly acquire America's higher death rates.[17]

If treatment has little impact and often comes too late, real improvements can only come by preventing the disease in the first place.

> 'The control of coronary heart disease necessarily depends on prevention, since treatment so often comes too late. Mass medication is potentially dangerous, and it would be better if risk factors could be controlled by changing habits.'

(British Medical Journal[18])

And the extremely low death rates in Japan, together with successful campaigns in other countries, show that heart disease is indeed largely preventable.[17]

Population studies and clinical research with volunteers have identified the main risks as faulty diet, smoking, overweight, too much alcohol, lack of exercise and excessive stress.[8] The Coronary Prevention Group points to a clear link between the consumption of animal fats and dairy products, and Britain's epidemic of heart disease.[19] Too much saturated fat in the diet can mean high blood cholesterol levels that in turn accelerate the build up of fatty deposits in the arteries. Doctors and the Health Education Council therefore recommend eating less meat and dairy products and suggest that saturated fats should be replaced by foods high in polyunsaturates such as certain vegetable oils. They also advise people to eat more fibre.[8]

Another factor is smoking, which doubles the risk of dying from heart disease.[20] Smoking, being overweight, drinking too much alcohol, lack of exercise, too much salt in the diet and excessive stress, all cause high blood pressure that again increases the risk of heart attacks,[8] not to mention strokes,[21] the third biggest killer in the United Kingdom.

North Karelia is the most easterly county of Finland and in 1971 had the highest death rate from coronary disease in the world. Not surprisingly the people were worried and set up a community programme. Everyone was advised to stop smoking, eat less fat and more vegetables, avoid obesity and have their blood pressure checked. By 1979, deaths from heart disease had fallen by 24 per cent in men and 51 per cent in women, a tremendous improvement over the rest of Finland.[5] In 1968 American men had one of the highest death rates from coronary heart disease, but thanks to a 56 per cent reduction in the consumption of animal fats, and a comparable rise in vegetable oils, American death rates fell by 21 per cent over the next 10 years.[22] According to the *British Medical Journal*'s ABC of Nutrition, 'Coronary heart disease is not an inevitable consequence of either ageing or industrialization. It appears to be largely preventable and the World Health Organisation recommends preventive community action in countries with a high incidence.'[5]

Inevitably, and despite the risks, mass medication has been suggested to lower cholesterol levels. Between 1964 and 1972 over 10,000 male volunteers were recruited to take part in a World Health Organisation trial using the drug clofibrate.[23] The outcome was alarming. Those receiving clofibrate did indeed suffer fewer mild, non-fatal heart attacks but the overall death rate turned out to be 37 per cent *higher* than those not taking the drug! As the *Lancet* put it, 'the treatment was successful but the patient died.' Although the idea of mass medication may be attractive to the drug industry, there will always be side effects (29 deaths in this case) and because of the dangers, dietary changes just have to be safer.

A similar picture emerges with cancer, the UK's second biggest killer.[3] It is difficult to imagine an area where resources have been so lavishly expended yet the disease shows no sign of decline.[24] In fact, since the early 1930s, the overall death rate has increased.[33] Earlier diagnosis may give the impression that cancer patients live longer but for the major types there have been few improvements.

According to the US Congress Office of Technology Assessment, improvements in the survival of people afflicted with solid tumours, those responsible for the vast majority of cancer deaths, have only been small with mortality rates for patients with cancers of the breast, stomach, colon, rectum and prostate hardly changing since 1950.[26] Breast cancer is the commonest cancer amongst British women and its incidence has increased from 335 deaths per million women in 1945 to 479 by 1979.[27] By 1983 it had risen again to 497 per million.[28] Reports indicate few advances in treatment so that life expectancy has hardly changed since 1945.[29] Lung cancer is the most common type amongst British men,[28] and also the most lethal,[30] followed by colorectal cancer. Now the second most common cancer in the western world,[30] colorectal cancer claims around 17,000 lives every year in the UK[31] and there has been little change in survival rate over the past 20 years.[30,32]

Table 1: Britain's major cancers for men (death rates greater than 100 per million)

Type	1971 – 75	1976 – 80	Percentage change
Bladder	118	123	+4.2
Pancreas	118	125	+5.9
Prostate	177	199	+12.4
Stomach	298	278	−6.7
Colorectal	311	320	+2.9
Lung, trachea, bronchus, etc.	1091	1125	+3.1

Table 2: Britain's major cancers for women (death rates greater than 100 per million)

Type	1971 – 75	1976 – 80	Percentage change
Pancreas	101	110	+8.9
Ovary, Fallopian tube, etc.	142	147	+3.5
Uterus	148	145	−2.0
Stomach	202	188	−6.9
Lung, trachea, bronchus, etc.	249	305	+22.5
Colorectal	352	355	+1.0
Breast	449	474	+5.6

Despite the massive rise in animal experiments over the last 30 years, death rates for the most common forms of the disease have hardly changed during the 1970s.[33] Tables 1 and 2 show how mortality rates for the major types of cancer either remained much the same or increased. Only cancer of the stomach and uterus declined but this is not due to improved treatment – fewer people are being afflicted with these diseases in the first place.[26] If treatment had really been effective, death rates would surely have declined significantly.

A recent article in the *New England Journal of Medicine* assessed progress against cancer in the United States during the years 1950 to 1982. Despite progress against some rare forms of cancer, particularly amongst patients under 30, accounting for 1 to 2 per cent of total deaths caused by the disease, the report found that the overall death rate had increased substantially since 1950: 'The main conclusion we draw is that some 35 years of intense effort focussed largely on improving treatment must be judged a qualified failure.' The report further concluded that '. . . we are losing the war against cancer' and argued for a shift in emphasis towards prevention if there is to be substantial progress.[76]

The cut, burn and poison approach to conventional cancer treatment can also have terrible side effects, In one study, patients suffering from breast cancer were asked to fill in questionnaires about their treatment.[34] Of patients, 42 per cent receiving a single drug and 79 per cent on multiple drug treatment said that side effects were severe enough to interfere with their lifestyle, whilst 29 per cent of patients receiving several drugs voluntarily added that treatment was 'unbearable' or 'could never be gone through again.' And because anticancer drugs are often carcinogenic themselves, there is a more than 5 per cent chance of contracting the disease again, even if the initial treatment is successful.[35]

All this may be dispiriting but it does highlight the desperate need for preventive action. Once again, careful population studies have identified the main causes with the result that cancer is now considered a largely preventable disease: '. . . 80-90 per cent of human cancer is determined environmentally and thus theoretically avoidable' (International Agency for Research in Cancer[36]) and 'It is now clear . . . that most, if not all cancers have environmental causes and can in principle be prevented' (Sir Richard Doll, Professor of Medicine, University of Oxford[37]).

'Environment' not only refers to the workplace but to eating habits and lifestyle factors. The main risks are connected with smoking and diet but also include excessive exposure to sunlight, alcohol, food additives, occupational hazards, pollution and even some drugs and medical procedures.[38] In Britain smoking accounts for a staggering 40 per cent of male cancer deaths[39] and worldwide is estimated to cause 900,000 new cases every year.[36]

As with smoking, diet is also thought to cause over a third of all cancers[40] but there is strong evidence that a vegetarian diet can be highly beneficial and protect against the disease. A massive cancer study in Japan, involving more than 100,000 people, revealed that individuals at highest risk are those who smoke, drink alcohol and eat meat, but who do not eat vegetables daily.[41] The safest group included those who did not smoke, drink or eat meat, but who did eat vegetables every day. This confirms earlier findings amongst the largely vegetarian Seventh Day Adventists who have cancer rates considerably lower than those for the general population.[42] The vegetarian diet also protects against breast and endometrial cancer in women[43], and perhaps even stomach cancer[44], whilst in Norway and America risks of colon cancer reduced by 25 per cent in men who ate large amounts of vegetables.[45] Finally, low-fat diets can protect against several cancers, particularly endometrium[45], and also breast and colon cancer.[47]

Obesity can also be a problem and a large American Cancer Society study found a close connection between the overweight and cancers of the endometrium, gall bladder, uterine cervix, colon and rectum, breast, ovary and other sites.[46]

The International Agency for Research on Cancer, established by the World Health Organisation, explains how many of the dietary factors implicated in cancer – obesity, too much meat and fat, lack of fibre – are characteristic of the Western diet[36] and, indeed, many cancers linked to such diets – for instance breast, colon and prostate – occur much more often in developed nations. Cancers of the colon and rectum are ten times more common in developed countries like the UK and America than in the Third World.[44]

Compared with smoking and diet, cancers caused by medical treatments are relatively small but still alarming as drugs are supposed to cure, or at least do no harm. According to an American survey[40] drugs and medical procedures cause between 2,000 and 12,000 cancer deaths every year in the United States. Examples include adenocarcinoma of the vagina in young women whose mothers had taken stilboestrol during pregnancy[37], brain tumours from drugs taken to prevent rejection of kidney transplants[37], and endometrial cancer after eostrogen replacement therapy,[48] particularly in women aged 45-74. Of all childhood cancers in Western Europe and North America in the 1950s and 1960s, 5-10 per cent were caused by X-rays used for diagnosis during the mother's pregnancy.[37] Ironically many anticancer drugs themselves can cause the disease.[35]

Estimates of cancer caused by pollution, occupational hazards and industrial products vary widely and a report for the US Congress Office of Technology Assessment, published in 1982,

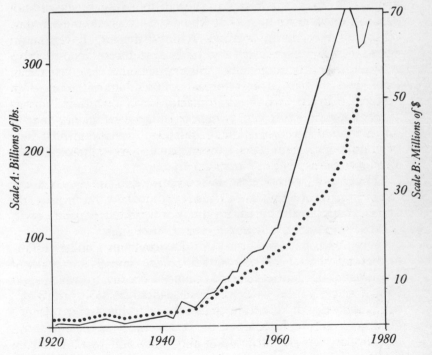

Key:
(—) *Synthetic organic chemicals,*
       Scale A
(....) *All chemicals and allied*
       products, Scale B

put the figure at 6 per cent with 15 per cent as an upper limit.[40] Examples include asbestos, the common industrial chemical benzene, and vinyl chloride in the manufacture of PVC.[37] But, even if reliable, a 6 per cent figure could be misleading about *future* rates. This is because the production of potentially hazardous industrial chemicals has only reached very high levels since the early 1960s (see graph on preceding page) and cancer sometimes takes up to 30 years to appear.[49]

To what extent these chemicals are dangerous is unknown. Of the 44,000 chemicals listed by the US Environmental Protection Agency as being in commercial production, only 7,000 have been subjected to any form of testing.[49] This may sound like an argument in favour of animal experiments to predict the risk of cancer but, apart from constant criticism of their scientific validity (Chapter 3), such tests take three years to perform and cost around £250,000 per chemical, so could never cope with the enormous number of substances. The urgent need for more sophisticated and especially quicker and cheaper tests led to the development of numerous *in vitro* or test tube techniques, which at least make it possible to tackle the problem.[50] In any case, despite experiments on animals, of those agents known to cause cancer in people, the great majority were *first* identified by observation of human populations whilst for several human carcinogens – for instance arsenic, benzene and alcohol – animal tests are persistently negative.[39]

Cancer and heart disease have been called the diseases of civilization, or of affluence, and indeed, much of our burden of disease can be attributed to the typical Western diet, and because we smoke, eat and drink to excess, fail to exercise properly and live and work in environments polluted by dangerous substances. A report by the National Advisory Committee on Nutrition Education shows that the typical British diet is a main cause of the diseases that most people in the West suffer and die from, including heart disease, cancer of the colon, obesity, high blood pressure, diabetes, constipation, diverticular disease, irritable bowel syndrome, gall bladder disease and strokes.[51] They recommended reduction in weight in the overweight and far less fat, sugar and salt in the diet as crucial steps towards preventing these diseases.

A third of all men and women in Britain are overweight[51] yet over 2,400 years ago[52] Hippocrates observed that, 'People who are naturally very fat are apt to die earlier than those who are

Britain's major killers (figures for 1985)

Cause of death	Number of people	Percentage of total deaths	Major risks
Ischaemic heart disease	163,104	28	diet, smoking, high blood pressure
Cancer	139,822	24	diet, smoking, alcohol, chemicals
Strokes	73,219	12	smoking, high blood pressure
Respiratory diseases *	64,607	11	smoking
Accidents	12,948	2	alcohol

* Mainly chronic bronchitis, obstructive lung disease and pneumonia
Source: OPCS Series DHI no. 17 (HMSO, 1987)

slender.' And modern life insurance statistics support his views: men who are more than 25 per cent above the average for their age and height have a death rate twice as high as those within 5 per cent.[2] The differences are due to an increase in heart disease, cancer, strokes, diabetes, kidney disease, gallstones, and accidents.[53]

Obesity has been closely linked with maturity onset diabetes[44] – the form of the disease that can usually be treated by diet alone.[54] In India, as long ago as 400 BC, it was observed that diabetes is a disease of the well fed and is now well known to be very much more common in the overweight.[44] The chances of becoming diabetic increase sharply as the level of obesity rises and in the UK during 1984 the disease claimed 6,369 lives.[3]

There has been a dramatic increase in deaths and diseases related to alcoholism,[55] with alcohol being a direct or contributory factor in an amazing 20-30 per cent of hospital admissions.[56] Not only that but alcohol abuse kills over 25,000 people every year in the UK alone.[57] Apart from cancer, alcohol plays a prominent part not only in deaths from cirrhosis, accidents and suicides, but in some deaths from cardiovascular, infective, respiratory and gastrointestinal conditions. It can also damage the foetus and may contribute to perinatal mortality.[58] Every year 10 per cent of deaths in people under 25 are caused by drunkenness. Alcohol is also linked with 80 per cent of deaths from fire, 65 per cent of serious head injuries, half of all murders, 40 per cent of road traffic accidents that include pedestrians, 25 per cent of fatal accidents, a third of all domestic accidents and 14 per cent of drownings.[59] In a third of all child abuse cases one parent drinks heavily.[59]

Between 1970 and 1978, deaths from alcoholism in the UK increased by 38 per cent in men and 130 per cent in women; deaths from cirrhosis of the liver increased by 36 per cent in men and 243 per cent in women whilst hospital admissions for the treatment of alcoholism and alcoholic psychosis increased by 77 per cent in men and 137 per cent in women.[55] England and Wales is now thought to have between 700,000 and 1,300,000 problem drinkers.[60]

Smoking has also been a traditional feature of affluent Western society but the recent massive sales drive in the Third World will undoubtedly kill millions more. Doctors consider the habit the single most preventable cause of death in the UK,[61] being responsible for 100,000 premature deaths a year.[62] Around 40 per cent of male cancer deaths are linked with cigarettes[39] and the impact on our health is enormous. The *British Medical Journal* states:

'. . . tobacco still accounts for 15-20 per cent of all deaths in Britain; of every 1,000 young men who smoke, one will be murdered, six will die on the roads, but 250 will be killed before their time by tobacco. Aside from cancers (of the lung, mouth, larynx, oesophagus, pancreas and urinary tract) and heart disease, smoking also kills thousands of patients with chronic bronchitis and emphysema and peptic ulcer – and it is an important avoidable hazard to the foetus.'[63]

Recent evidence now links cigarettes with leukaemia[64] whilst the Royal College of Surgeons estimate that at least 90 per cent of deaths from lung cancer and two of the three main respiratory diseases – chronic bronchitis and obstructive lung disease – are attributable to smoking.[65] And in the Framingham study, where the inhabitants of a small town in Massachusetts voluntarily supplied information about their lifestyle and health status, heavy smokers were found to have three times the risk of a stroke.[66]

But, apart from the diseases of affluence, there are striking differences between social groups that indicate that the direct and indirect effects of poverty, such as bad housing, are still major factors in disease. Working-class people suffer much higher levels of chronic sickness and death rates than do employers and managers.[67] The differences are most striking in children under one (see graphs overleaf) but can also be seen in adult life, particularly with cancers, infections, diseases of the nervous system and sense organs, blood disorders, diseases of the respiratory, digestive and genito-urinary systems, and accidents and violence.[67] One example, as you may recall from Chapter 1, is Professor Stewart's finding that a child's social class is three times more important than vaccination in influencing whooping cough outbreaks.

In 1977 the Labour government set up a working group chaired by the then Chief Scientist at the DHSS, Sir Douglas Black, to investigate the inequalities in health. The Secretary of State for Social Services said at the time:

> '... in 1971 the death rate for adult men in social class V (unskilled workers) was nearly twice that of adult men in social class I (professional workers), even when account has been taken of the different age structure of the two classes. When you look at death rates for specific diseases the gap is even wider. For example, for tuberculosis the death rate in social class V is ten times that for social class I; for bronchitis it was five times as high and for lung cancer and stomach cancer three times as high. Social class differences in mortality begin at birth. In 1971 neonatal death rates – deaths within the first month of life – were twice as high for the children of fathers in social class V as they were in social class I. Death rates for the post-neonatal period – from one month up to one year – were nearly five times higher in social class V than in social class I.'[68]

In 1980 the Committee published its findings. Not only had there been little improvement but the situation had actually got worse during the 1960s and early 1970s. The report pointed to poverty, poor living and working conditions and general deprivation as major causes of ill health. One survey found that 29 per cent of preschool children in the lower social classes, compared with 3 per cent in social class I, lived in inadequate

Infant mortality rates (less than one year old)

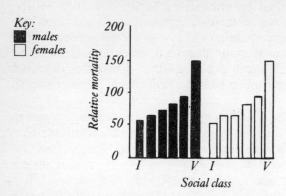

Diseases of the respiratory system

Infective and parasitic diseases

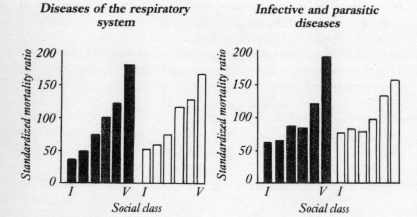

Malignant neoplasms

Diseases of the genito-urinary system

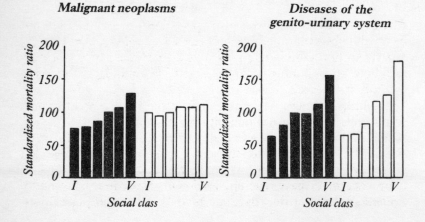

accommodation. The report made many recommendations including an enlarged programme of health education by the government, improvements in living and working conditions and, particularly, the abolition of child poverty. Success would bring dramatic results: if the death rates in social class I had applied to classes IV and V during 1970-72, 74,000 lives would have been saved, including 10,000 children and 32,000 men of working age.[67]

Six years after the Black Report, new evidence for the period 1979-1983 shows that the health gap had widened still further, as more people live at or near the poverty line.[69] So, despite the tremendous efforts of the social reformers in the nineteenth century, much still remains to be done.

As we have seen, the major causes of sickness and death in our society include heart disease, cancer, strokes, respiratory diseases, alcohol-related problems, high blood pressure and diabetes, most of which are very difficult or impossible to cure, but all largely preventable. If major advances in our health are to be achieved the emphasis, as in the past, must once again be on prevention. Just as social reformer Lemual Shattuck said in 1850 when referring to the infectious epidemics, '. . . measures for prevention will affect infinitely more than remedies for the cure of disease'[70] And Chadwick's general conclusion still applies: by tackling causes rather than effects, it will cost society less to prevent disease than to cope with its victims. Over 300 years ago Thomas Adams said it all:

> 'Hee is a better physician that keepes diseases off us, than hee
> that cures them being on us. Prevention is so much better
> than healing, because it saves the labour of being sick.'[36]

And since the principal causes of these diseases were discovered by careful observation of human behaviour, it must follow that *at best* vivisection could only play a small part in improving our health, even assuming it to be a reliable or indeed the only method of research.

Naturally enough we have concentrated on the diseases responsible for the great majority of deaths in the UK but the same *principle* applies to other, less common disorders. Priority should be given to find the cause so, whenever possible, preventive action can be taken. In the case of AIDS, action to prevent the disease reaching epidemic proportions is urgently needed. Soon after the first reports in 1981, careful population

studies indicated that AIDS is a sexually transmitted disease
mainly affecting the gay communities both here and in the
United States. But the disease is not confined to homosexuals
and can affect anyone who is sexually active – including, as the
British Medical Journal put it, '. . . the "innocent" partners of the
promiscuous' and drug addicts who share needles.[71] These
observations suggested the means of preventing further spread
of the disease: by changing sexual practices and taking
precautions with injection needles and blood transfusions.
Already, vigorous health education campaigns in San Francisco
have proved successful in changing behaviour and slowing the
spread of the disease.[71] Practical measures such as the use of
condoms have also helped slow the spread of AIDS amongst
those at very high risk – for instance prostitutes.[104]

A special report by the United States National Academy of
Sciences entitled 'Confronting AIDS' argues that the most
effective measures for reducing the spread of AIDS are
education and voluntary changes in behaviour. Even if a vaccine
or drug does become available, the report concludes, '. . . public
education about HIV infection (AIDS) is, and will continue to
be, a critical public health measure.'[105] This has been proven by
other sexually transmitted diseases that, in the UK, had virtually
doubled from 338,000 new cases in 1971 to 658,000 by 1985
despite the availability of modern drugs.[107] So prevention is
better than cure even when effective treatment is available, but
as the *BMJ* explained, for AIDS '. . . prevention is not just better
than cure – it is the *only* cure.'[71]

It will come as no surprise that medical scientists are working
to develop a vaccine against AIDS, but there could be major
problems. Vaccines against other viruses mobilize the body's
immune system. No vaccine has ever been developed against a
virus that actually attacks the body's natural defences. The
United States National Academy of Sciences report states:

> 'Developing a vaccine to prevent HIV infection and AIDS
> presents a number of scientific challenges that have never
> before been responded to successfully. As a result, an
> effective vaccine may be very difficult, if not impossible, to
> produce. Should an effective vaccine candidate become
> available, there are significant social concerns that may limit
> or prevent its testing and use . . . Even for the next five to ten
> years, the committee generally believes that the probability of
> a licensed vaccine becoming available is low.'[105]

But no one can afford to wait years for a vaccine even it it does prove safe and effective. Vigorous and explicit health education campaigns need to be initiated *now* to stop the disease spreading. Like other sexually transmitted diseases, AIDS can only be effectively controlled by prevention, through voluntary changes in behaviour, and no amount of animal experimentation can achieve that.

The technique that so often provides the evidence on which to base disease prevention campaigns is called epidemiology. Scientists study entire populations or smaller sub-groups and link diseases with lifestyle or environmental factors. The relationship between smoking and lung cancer was first discovered in this way[37] and risk factors associated with heart disease – smoking, high blood pressure, lack of exercise, being overweight and excess cholesterol – were all discovered by epidemiology.[73] For instance Belgium turns out to be an excellent 'laboratory' for testing dietary theories because of its division into two distinct populations – the French speaking southerners, who eat four times as much butter as the Dutch speaking northerners, who have lower cholesterol levels and fewer deaths from heart disease.[22] Often epidemiology will suggest clues to be followed up by clinical case studies with much smaller numbers of people.

Before 1914 epidemiology had identified several causes of cancer.[39] It was found that pipe smokers were more prone to lip cancer and radiologists had an increased risk of skin cancer. Workers in the aniline dye industry contracted bladder cancer and patients painted with coal tar ointment developed cancer of the skin. Unfortunately the scientific world then became infatuated with laboratory experiments in which cancers were artificially induced in animals and epidemiology (the search for underlying causes) lost favour.

'By this time (1918) Yamagiwa and Ishikawa had produced cancer on a rabbit's ear by painting it with tar, and the potential for laboratory experiments had captured the imagination of the scientific world. Observational data, which had not been obtained by experiment, required an element of subjective interpretation and were commonly dismissed on the grounds that conclusions drawn from them were uncertain and could be misleading. They carried little weight in comparison with those obtained by experiments, and there was little point in worrying about them, as it was confidently

believed that the mechanism by which all cancers were caused
would be discovered within a few years.'
                                                    (Sir Richard Doll, 1980[39])

These hopes were to be disappointed and eventually interest in
epidemiology was rekindled as scientists became more concerned about the causes of the disease. As a result cancer is now
known to be largely preventable.

Although the vivisectors had their own 'tools of research',
they were not needed to identify the causes of the UK's
twentieth century epidemics – the diseases of civilization. At
best they could confirm what was already known from
epidemiology or clinical observation of patients, but often served
only to confuse the issue. For example, attempts to duplicate the
carcinogenic effect of cigarettes by forcing animals to inhale the
smoke, have proved 'difficult or impossible' according to the
Lancet.[74] Furthermore the important US Congress Office of
Technology Assessment report on the causes of cancer, relied
far more on epidemiology than on animal experiments or other
laboratory studies because, its authors argued, these '. . . cannot
provide reliable risk assessments.'[46]

Suspicions that vivisection has little impact are confirmed by
recent analysis of trends in diseases which indicate that our
overall health could actually be deteriorating despite the large
increase in animal experiments since 1950. The General
Household Survey reveals a progressive *increase* in the number
of people reporting chronic sickness[79] (defined as any long
standing illness, disability or infirmity) between 1972 and 1982.
In 1972 29 per cent of men and 31 per cent of women aged
45-64 reported[99] being chronically sick but this had increased to
41 per cent and 42 per cent respectively by 1982.[78] The trend is
confirmed by government figures which show an increase in the
number of insured people incapacitated by illness.[80] Levels of
chronic sickness have also increased in the elderly which
suggests that although people are living a little longer, their
quality of life may not have improved. Certainly evidence from
America suggests that middle-aged people can expect more
years of disability and fewer years of active life than they could a
decade before.[81]

Further indications come from hospitalization rates which
have increased from 78 per 1,000 in 1958 to 103 per 1,000 in
1974.[81] According to the Department of Health and Social

Security, hospital admissions are rising by 1.5 to 2 per cent a year.[52] As if to reflect the increasing level of chronic sickness, the number of prescriptions issued per person has risen from an average of 4.7 per person in 1961 to 7.0 in 1985.[107] In fact 393.1 million prescriptions were dispensed during 1985, a dramatic increase over the 233.2 million issued in 1961.[107] And when it comes to pneumonia the picture for people over the age of 65 is grim. Since the turn of the century death rates fell rapidly (as mentioned in Chapter 1), but from the mid 1940s, when antibiotics became available, the rate for people aged 65-84 has risen sharply.[82] Our elderly population seem to be more vulnerable and less able to withstand diseases like pneumonia, despite the availability of modern drug therapy. It does not look as if we are getting healthier!

The epidemic of heart disease, the mounting toll of cancer victims, the increasing level of chronic sickness and the widening health gap between the social classes are all crippling the National Health Service, which simply cannot cope. Drugs have not been able to solve these problems, so if the nation's health were a government priority we could expect a vigorous campaign to prevent disease. Yet, the reality of government health policy is that less than 0.4 per cent of the nation's health budget goes on specific preventive measures![108]

By 1983 Britain had the unenviable distinction of being top of the world's international league table for heart disease, death rates having remained much the same since the early 1970s.[83] Yet, despite over 150,000 largely preventable deaths, government not only fails to take effective action but actually increases expenditure on the heart transplant programme.[84] Transplants may bring sensational headlines and a few extra votes but even if they ever approached a 100 per cent success rate, could never make any significant impact on the enormous toll of premature deaths.

When Sir Douglas Black presented his Report on Inequalities in Health, with proposals to reduce the health gap between social classes and save tens of thousands of lives, the government would not endorse its recommendations because, they said, it would cost too much.[67] Not even a start could be made!

It is true that government has recently launched a campaign to reduce drug abuse and there is no doubt that the problem needs to be tackled. But consider the figures.[57] In 1984 drug abuse led to 235 deaths and rightly caused great concern. Yet in the same

year over 100,000 people died prematurely through smoking
and over 25,000 through alcohol abuse. The British government
spent £22 million trying to combat illicit drug abuse in 1984 but
could only manage £150,000 grant to Action on Smoking and
Health (ASH) and refused to give anything at all to Action on
Alcohol Abuse.[57]

Why has the government so little interest in prevention?
Could it be that any action towards a healthier lifestyle would
inevitably conflict with powerful vested interest groups such as
the tobacco and alcohol industries, the meat and dairy industry,
the processed food industry and the drug industry whose
aggressive advertising and promotional activities encourage the
mistaken belief that health can be obtained from a bottle? Take
Britain's diet as an example. In 1983 the National Advisory
Committee on Nutrition Education (NACNE) announced its
recommendations for a healthier diet – far less fat, sugar and salt
as crucial steps towards preventing many of the UK's most
common fatal diseases. But, according to an investigation by the
Sunday Times, the report was suppressed by the DHSS for two
years and only then released with more modest
recommendations.[51] NACNE's proposals, if implemented,
could have severe economic consequences for the dairy industry
(a major producer of fat) and the food processing industry,
which makes heavy use of all three ingredients. Most of the
opposition to the NACNE report, it is revealed, came from the
British Nutrition Foundation, which receives 98 per cent of its
funding from the food industry. And the DHSS is reluctant to
act because of hostility from the Ministry of Agriculture,
Fisheries and Food, which speaks in Whitehall for the meat and
dairy industry and the food manufacturers. Subsequently the
Joint Advisory Council on Nutrition Education presented far
weaker proposals but, according to the *Lancet*:

> '. . . even this report, it seems, was received with suspicion by
> the Department of Health and Social Security who delayed
> publication in deference to the Ministry of Agriculture,
> Fisheries and Food, who are concerned about the health of
> the dairy industry. The Ministry of Agriculture, it seems, is
> also impeding the introduction of comprehensive food
> labelling, which would enable the consumer to learn how
> much to reduce intake of fat, sugar, salt and preservatives.'[85]

And groups like the Meat Promotion Executive, the Butter
Information Council and the Milk Marketing Board spend

millions to convince us about the goodness in meat and dairy products.[25]

In recent years the Butter Information Council and the National Dairy Council are reported to have conducted massive advertising campaigns to persuade doctors to reject the view of medical groups calling for a reduction in saturated fat.[86]

Despite constant appeals from health groups, government persistently refuses to take effective action against the tobacco industry and spends the equivalent of just one per cent of the industry's promotional budget on antismoking campaigns.[87] In 1979 the then Under Secretary of State at the DHSS, Sir George Young, stated at the Fourth World Congress on Smoking and Health:

> 'The solution of many of today's medical problems will not be found in the research laboratories of our hospitals but in our parliament.'[88]

Sir George, a committed antismoker, was subsequently removed from the DHSS, the department responsible for dealing with the tobacco industry, so effective action to eliminate cigarette advertising was blocked.[89] But the established medical journals could not be dismissed and continued to stress the facts.

> 'However boring it may be to reiterate plain facts, the truth is that smoking is still by far and away the largest preventable cause of death in Britain. Tragically it will remain so as long as governments fail to act on what we have known for years.'
> (*British Medical Journal*[89])

Whilst Britain refuses to act, other countries have taken positive action, with encouraging results. In Sweden and Norway, severe government measures against tobacco advertising have resulted in a sharp drop in the numbers of people smoking.[90] In 1975 the Norwegian Tobacco Act banned all forms of promotion and stepped up health education about smoking.[90,91] But in Britain the government turns out to be a powerful vested interest itself, receiving £4.5 billion in tax revenue every year.[92] And early adult deaths save billions in unpaid pensions. So it is not surprising that information coming to light under the 30 year rule on access to government documents, reveals that the government of the day disregarded scientific evidence linking smoking to lung cancer in order to maintain its highly profitable tobacco tax.[93]

In an attempt to improve its image and persuade governments to continue the voluntary advertising arrangements, the

Tobacco Advisory Council, which speaks for all the British manufacturers, has set up the Health Promotion Research Trust.[94] This will investigate ways of improving the public health *with the exception of research into the effects of smoking*! In fact the Tobacco Advisory Council still does not accept the lethal nature of its products,

> 'We in the tobacco industry do not accept that there is any causal connection between smoking and the so-called smoking-related diseases.'[92]

Another subtle move, and one involving many animal experiments, is the search for a 'safe' cigarette, even though the industry appears to believe its products are harmless. Such an approach has been dismissed as 'futile' by the *BMJ*[63], but is no doubt intended to divert attention from the real issue – that advances must come from government action rather than laboratory experimentation, as Sir George Young pointed out in 1979. In any case, low-tar cigarettes are still dangerous and there seems to be no evidence that cigarettes of reduced tar or nicotine yield reduce deaths from heart disease – the greatest killer amongst the smoking-related diseases.[63] The few that do smoke low-tar, low-nicotine cigarettes (15 per cent of British smokers) compensate by smoking more intensively, by puffing and inhaling harder.[62]

Richard Peto, an expert in cancer studies at the University of Oxford, argues that the most powerful pressure groups in the United States have been the large financial interests that '. . . have always put financial advantage before human health.' In the science magazine *Nature* he states:

> 'No other industry kills people on anything like the scale that the tobacco industry does, but where other industries have been found to cause cancer (or dust-induced lung diseases) in their workers or in the consumers of their products, their immediate response has usually been to delay acceptance of the findings, to minimize their relevance to current practice, and in general to delay or obstruct any hygienic measures that will cost money.
>
> 'Large amounts of money are available to mount press and TV publicity campaigns about the homely apple-pie virtues of asbestos, journals financed by the tobacco industry run populist articles that misrepresent research results to lay readers, and some US television journalists are explicitly told always to censor all reference to the dangers of smoking.'[47]

Like tobacco, alcohol sales are also highly profitable for the government, bringing in a massive £5.8 billion in taxes and excise duty,[95] together with £900 million in exports.[59]

> '... no wonder successive governments have suppressed reports calling for curbs on advertising and realistic price rises at home and put pressure on the Council of Europe and the World Health Organisation to modify initiatives that would affect Britain.'   (Editorial, *British Medical Journal*[96])

Apart from the massive annual income, Dr Stuart Horner, the chairman of the Institute of Alcohol Studies, has accused the government of having vested interests in the drinks trade.[97] According to Dr Horner, 11 government ministers had direct financial links with brewers and distillers before taking office and a further 50 MPs had direct or indirect financial interests in the drinks trade.

Another powerful interest group is the pharmaceutical industry, whose primary purpose is to make a profit and satisfy its shareholders. The industry is unlikely to be interested in prevention unless it can be made to pay, for instance in the case of vaccines. It is naturally more interested in treating the symptoms of disease rather than preventing it in the first place. Without widespread sickness, the industry cannot thrive. Studies have shown that 80 per cent of people who seek medical help have conditions that either would get better if nothing were done or cannot be improved by drug treatment.[75] This is because most disease is self-limiting – the body being able to cure itself if given the chance. But heavy promotion by the industry helps ensure that doctors usually reach for the prescription pad as the fast formula for health. As a result most people (77 per cent) consulting their GP obtain a prescription.[106] Inevitably patients come to adopt the 'pill for every ill' mentality and the idea that health can be obtained from a bottle. As Health Action International explains:

> 'Not only do companies sell their particular branded drug, but indirectly they sell the concept that there is a "pill for every ill". As most drugs are administered after an illness has set in, they also reinforce a focus on curative rather than preventive medicine and lead to a prevalence of what one doctor has called "symptom swatting" rather than careful identification of causes of illness.'[77]

As the number of prescriptions has soared, iatrogenic or

drug-induced disease has reached what DHSS Principal
Medical Officer, R. D. Mann[98], calls 'epidemic proportions.'
Medical reports reveal that 3-8 per cent of hospital admissions
are caused by the adverse effects of drugs whilst as many as 40
per cent of patients may experience side-effects in general
practice.

The limited interest in preventive education extends to
leading medical research charities. The idea that cancer and
heart disease are largely preventable would hardly encourage
donations and receives relatively little publicity. In a recent
review of its work, the British Heart Foundation gave the
impression that *only more research* can make an impact on
Britain's epidemic of heart disease.

> 'The only certainty is, that without continuing research, we
> shall never make any advance in eradicating heart disease.'
> (British Heart Foundation, Annual Review[4])

During 1986/87, the Foundation allocated just four per cent of
its charitable expenditure to education, although only a part of
this was devoted to primary prevention.[109]

Overall Britain's cancer research charities receive over
£1 million *per week* in public donations,[72] yet the two main
bodies spend little on prevention. In 1987 the Cancer Research
Campaign (CRC) spent just 1.6 per cent of its total expenditure
on cancer education[100] whilst the Imperial Cancer Research
Fund (ICRF) appears to spend virtually nothing on preventive
cancer education,[101] despite assets of over £65 million.[72] An
editorial in the *BMJ* expressed its concern that the British heart
and cancer charities '. . . pay so little attention to prevention' and
confirmed that, for the CRC and ICRF in particular, prevention
plays only a '. . . minute part of their activities.' The two
charities, it says, '. . . give virtually no direct support to
preventive cancer education.'[1]

Attitudes towards prevention were revealed when the British
Medical Association published its explosive Report on Investment in the UK Tobacco Industry.[102] Numerous medical
charities and health authorities were found to have financial
holdings in the tobacco industry. Although smoking is a
well-known cause of cancer and heart disease, the BMA found
that the British Heart Foundation, the Imperial Cancer
Research Fund, the Institute of Cancer Research and the
Medical Research Council, which doles out over £100 million of

tax payer's money every year for research, all had substantial holdings in Grand Metropolitan, a company with considerable tobacco interests. In 1982/83 The Cancer Research Campaign had holdings in both Rothmans and British American Tobacco Industries, but had sold these by 1984.

A hundred years ago the social and humanitarian reformers had to battle against apathy and vested interests to achieve the necessary improvements in public health. Today's pressure groups will have to fight just as hard to overcome the financial interests of both industry and government that conflict with the nation's health. Not only that but in our culture treating disease is enormously profitable, *preventing* it is not. In 1985 the US, Western Europe and Japanese market in cancer therapies was estimated at over £3.2 billion with the 'market' showing a steady annual rise of 10 per cent over the past five years.[103] Preventing the disease benefits no one except the patient. Just as the drug industry thrives on the 'pill for every ill' mentality, so many of the leading medical charities are financially sustained by the dream of a miracle cure, just around the corner. In an article by the International Agency for Research in Cancer entitled 'The World Cancer Burden: prevent or perish' it is stated

> 'To promise substantial reduction after ten years [by prevention] is somehow not as appealing as that elusive will o' the wisp – the cure for cancer.'[36]

Common sense dictates that prevention is always better than cure but, however successful society becomes in preventing disease, there still has to be a place for properly conducted medical research, to help those who do become sick. The question now remains: are animal experiments the best way to understand disease and its treatment and do they provide adequate protection against hazardous drugs?

1 *BMJ*, 1610, 22-29 December, 1979
2 T. McKeown, *Role of Medicine* (Blackwell, 1979)
3 Office of Population Consensus and Surveys, *Monitor*, 1984, DH2 85/3 (HMSO, 1985)
4 British Heart Foundation Annual Review, 1980/81
5 A. Stewart-Truswell, *BMJ*, 34-37, 6 July, 1985
6 B. Lewis, *BMJ*, 177-180, 19 July, 1980
7 C. Wood, *New Scientist*, 656-658, 3 June, 1982
8 'Beating Heart Disease', Health Education Council
9 H. G. Mather et al, *BMJ*, 334-338 7 August, 1971
10 A review by B. Lewis in the *BMJ*, ('Dietary prevention of ischaemic heart

disease – a policy for the 80s' – reference 6) puts the contribution of coronary bypass surgery in reducing overall mortality at 4-5 per cent 'on generous assumptions.' Other *BMJ* correspondents also believe its impact is small, for example, A. M. Johnson, 1691, 15 December, 1984 and P. Ford, 1690-1691, 15 December, 1984. An editorial in the *New England Journal of Medicine* (E. Braunwald, 663, 22 September, 1977) stated: 'An increasing number of patients are being operated upon ... because of the hope, largely without objective supporting evidence at present, that coronary artery bypass graft prolongs life or diminishes the frequency of subsequent myocardial infarction (heart attack).'

11 M. L. Murphy, et al, *New England Journal of Medicine*, 621-627, volume 297, 1977
12 Reported in *Health Shock* by M. Weitz and references therein (David & Charles, 1980)
13 D. Gould, *The Times*, 7 December, 1979
14 *Journal of the American Medical Association*, 1616, volume 238, 1977
15 'Costs and benefits of the heart transplant programme at Harefield and Papworth Hospitals', DHSS (HMSO, 1985)
16 *Lancet*, 356-7, 12 August, 1978
17 G. Rose, *BMJ*, 1847-1851, 6 June, 1981
18 *BMJ*, 747-750, 15 March, 1980
19 *The Listener*, 10, 3 July, 1980 based on the BBC2 Brass Tacks programme, 'Full of Natural Goodness'.
20 *BMJ*, 573, 30 August, 1980
21 *Lancet*, 1195–6, 28 May, 1983
22 *World Medicine*, 7, 31 May, 1980
23 *Lancet*, 1131-2, 25 November, 1978
24 *BMJ*, 1732, 12 June, 1982. Tony Smith, deputy editor, states that 'the incidence of the disease in Western countries shows no signs of decline.'
25 S. Campbell, *Meat Industry*, October 1983 and reference 19
26 M. Gough and H. Gelband, *Nature*, 598, volume 296, 1982
27 *Hansard*, 472, 3 November, 1980
28 OPCS, 1983, DH2 no. 10 (HMSO, 1984)
29 In *Hansard* (10 November, 1980) the DHSS stated: 'There is no evidence that the average life-expectancy for women with breast cancer in this country has changed significantly since 1945.' A *Lancet* editorial (10 October, 1981) comments: 'Breast cancer remains a common and often fatal disease and the evidence that developments in its treatment have had a favourable effect on the duration or quality of survival remains disappointing.'
30 J. D. Hardcastle et al, *Lancet*, 1-4, 2 July, 1983
31 *Lancet*, 996, 27 April, 1985
32 J. D. Hardcastle, *Lancet*, 791-793, 12 April, 1980
33 OPCS DH1 no. 15
34 B. V. Palmer, et al, *BMJ*, 1594-1597, 13 December, 1980
35 *Lancet*, 261-2, 4 February, 1984
36 C. S. Muir and D. M. Parkin, *BMJ*, 5-6, 5 January, 1985
37 R. Doll, *Nature*, 589-596, 17 February, 1977
38 For example, R. Doll, *Cancer*, 2475-2485, volume 45, 1980; R. Doll, *Nutrition & Cancer*, 35-43, volume 1, 1979; *Newsweek*, 40-45, 26 January,

1976; and reference 37

39 R. Doll, *Cancer* 2475-2485, volume 45, 1980

40 US Congress Office of Technology Assessment report on the causes of cancer by R. Peto and R. Doll reported in *World Medicine*, 33, 21 August, 1982

41 Reported in *New Scientist*, 8, 15 November, 1984

42 R. L. Phillips, *Cancer Research*, 3513-3522, volume 35, 1975

43 *BMJ*, 23 January, 1982

44 The *BMJ*'s ABC of Nutrition (20 July, 1985) notes that salads and citrus fruits protect against stomach cancer, whilst dried, salted fish, pickled foods and cured meats are among those factors that cause the disease. The International Agency for Research on Cancer also note that diets rich in fresh fruit and vegetables can protect against stomach cancer (reference 36)

45 R. Doll, *Nutrition & Cancer*, 35-43, volume 1, 1979

46 F. J. C. Roe, *BMJ*, 1421-1422, 28 November, 1981

47 R. Peto, *Nature*, 297-300, 27 March, 1980

48 *Lancet*, 1359-1360, 20 June, 1981

49 D. L. Davis and B. H. Magee, *Science*, 1356-1358, volume 206, 1979

50 These tests are discussed in Chapter 6 and so the references are listed there.

51 Reported in *New Scientist*, 114-115, 14 July, 1983

52 *Prevention and health: everybody's business*, DHSS, 1976 (HMSO)

53 Some of the diseases associated with obesity are listed by the Health Education Council's booklet 'Looking after Yourself'; see also references 2, 44 and 46

54 The main forms of diabetes are a) juvenile onset diabetes: this usually requires insulin and affects about 10 per cent of diabetics and b) maturity onset diabetes, which affects about 85 per cent of diabetics. This form can usually be treated by diet alone. A study published in *Diabetes*, 789-830, volume 19, 1970, showed that, in terms of survival, there was little difference between insulin and diet alone for this form of the disease.

55 *BMJ*, 464, 9 August, 1980

56 *Lancet*, 117-8, 22 November, 1980; *BMJ*, 502, 14 February, 1981

57 *BMJ*, 712-3, 15 March, 1986

58 R. Smith, *BMJ*, 835-8, 26 September, 1981

59 *BMJ*, 779-780, 17 September, 1983

60 R. Smith, *BMJ*, 1043-1045, 17 October, 1981

61 *BMJ*, 1281, 14 November, 1981

62 Report from the Royal College of Physicians entitled 'Health or Smoking?', described in *British Journal of Addiction*, 241-3, volume 79, 1984

63 *BMJ*, 1570-1, 26 November, 1984

64 *New Scientist*, 7, 7 March, 1985. This discusses 'passive smoking', that is, nonsmokers inhaling smoke from smokers in the household. Leukaemia appears seven times more often among people who have spent their lives with smokers

65 Reported in *Hansard*, 665-666, 14 December, 1984

66 *BMJ*, 573-5, 30 August, 1980

67 Inequalities in Health – report of a research working group, DHSS, 1980

68 Reproduced in reference 67
69 *Guardian*, 30 July, 1986
70 See Chapter 1
71 *BMJ*, 348, 9 August, 1986
72 Handbook of the Association of Medical Research Charities 1987/88
73 W. W. Holland and A. H. Wainwright in *Epidemiologic Reviews*, Ed.
   P. E. Sartwell, volume 1 (Johns Hopkins University Press, 1979)
74 *Lancet*, 1348-9, 25 June, 1977
75 The Wellcome Museum of the History of Medicine (Science Museum,
   London, 1986)
76 J. C. Bailar and E. M. Smith, *New England Journal of Medicine*, 1226-32,
   volume 314, 1986
77 Health Action International 'Problem Drugs' Pack, 13 May, 1986
78 *Social Trends*, no. 15, 1985
79 *Social Trends*, no. 14, 1984
80 *Social Security Statistics*, 1984, page 26 (DHSS)
81 A. Melville and C. Johnson, *Cured to Death* (New English Library, 1983)
82 According to the Registrar General's Statistical Review (1949) the death
   rate from pneumonia for people over 65 declined from 1931 to 1945.
   OPCS Trends in Respiratory Mortality 1951-1975 (DH1 no. 7, 1981)
   reveal a large increase in the pneumonia death rate for people aged 65-84
83 K. Ball, *Lancet*, 339-340, 6 August, 1983
84 *The Guardian*, 17 January, 1986
85 *Lancet*, 401, 17 August, 1985
86 K. Ball, *Lancet*, 1182, 1 December, 1979
87 *Nature*, 801, 28 February, 1980
88 Reproduced in reference 86
89 *BMJ*, 1281, 14 November, 1981
90 *Sunday Times*, 8 February, 1981
91 *BMJ*, 1800, 30 May, 1981
92 *Nature*, 418, 1 December, 1983
93 *Lancet*, 120, 12 January, 1985
94 *Lancet*, 409, 18 February, 1984
95 *BMJ*, 561-2, 16 February, 1985
96 A. Paton, *BMJ*, 1-2, 5 January, 1985
97 *Lancet*, 240, 28 July, 1984
98 R. D. Mann, *Modern Drug Use* (MTP Press Ltd, 1984)
99 *Social Trends*, volume 6, 1975
100 Cancer Research Campaign, annual report 1986, Handbook, 1987
101 Imperial Cancer Research Fund, annual report and accounts, 1985/86
102 Report on Investment in the UK Tobacco Industry, British Medical
    Association, 1985
103 *SCRIP*, 21, 20 May, 1985
104 G. L. Smith and K. F. Smith, *Lancet*, 1392, 13 December, 1986
105 *Confronting AIDS*, Institute of Medicine, National Academy of Sciences,
    1986 (Washington DC: National Academy Press)
106 *Social Trends*, no. 16, 1986
107 *Social Trends*, no. 17, 1987
108 D. Sanders, *The Struggle for Health* (Macmillan, 1985)
109 British Heart Foundation, annual report 1986/87

PART 2

Vivisection – The Myth

CHAPTER 3

Tools of research

'The idea, as I understand it, is that fundamental truths are revealed in laboratory experiments on lower animals and are then applied to the problems of the sick patient. Having been myself trained as a physiologist, I feel in a way competent to assess such a claim. It is plain nonsense.'

(Sir George Pickering[1]
Professor of Medicine, University
of Oxford)

In 1980 Eli Lilly introduced their new arthritis drug Opren. It had safely negotiated animal tests and was ready to take the market by storm. For Opren was no ordinary drug. It had a distinct advantage over its 23 rivals, which could only alleviate symptoms at best. Opren could do that but by appearing to modify the disease process it actually seemed to hold out promise of a cure – an enormous advantage that would guarantee its makers a huge slice of the UK's £100 million market in antiarthritis pills. The drug was aggressively promoted not only to the medical profession but to the public as well, ensuring that arthritis sufferers would soon be demanding Opren from their doctors. And it is true, Opren did indeed modify the disease process – but only in laboratory rats.[2] By the time human trials had disproved the animal data, it was too late. Opren turned out to be highly toxic with 3,500 reports of harmful effects including 61 deaths in Britain alone.[3] By August 1982, 22 months after its spectacular launch, Opren was withdrawn.

The idea that animal experiments can be so dangerously misleading is not widely advertised by the scientific community. Yet, there are countless known examples where animal tests have produced conflicting results: morphine sedates man but stimulates cats;[4] aspirin causes birth defects in rats and mice but not in people;[5] thalidomide works the other way around;[6] penicillin is highly toxic to guinea pigs and hamsters;[6] the common industrial chemical benzene causes leukaemia in man but not mice;[7] insulin produces deformities in laboratory animals but not in people;[8] nitrophenol causes cataracts in humans, ducks and chicks but not other laboratory animals;[9] serotonin, a naturally occurring chemical in the body, raises the blood pressure in dogs but reduces it in cats;[10] and doses of aspirin used in human therapeutics actually poison cats whilst having no effect on the treatment of fever in horses.[10]

Since animals can react differently to human beings, new treatments that are tested on animals must be tried out *again* in healthy volunteers and patients, before they can be considered safe and effective. By producing incorrect results animal experiments not only prove dangerous but actually have the effect of retarding clinical investigation – the only truly valid approach. This has serious implications for patients because animals are not only used to assess the value and safety of drugs, but to understand human disease, to develop surgical techniques and to acquire physiological knowledge.

Nor can anyone claim this is a new phenomenon, recently discovered with the growth of animal tests. Galen (AD 131-201) has been described as the founder of experimental physiology and indeed vivisected many animals. Inevitably he made some very serious mistakes because he assumed that people and animals were alike in the way their bodies worked. In his *History of Medicine*, D. Guthrie states that Galen's anatomy

'. . . was based mainly on the study of apes and pigs, and he unhesitatingly transferred his discoveries to human anatomy, thus perpetuating many errors.'[11]

Unfortunately Galen's dogmatic style, together with the Church's reluctance to allow dissection of human cadavers, meant his errors went uncorrected for literally hundreds of years.[12] Galen's mistakes thus passed into current teaching and became authoritative statements of the Universities until as late as the middle of the sixteenth century! So it's not surprising that medicine made such little progress through the Dark Ages.

Hippocrates' fine example, that disease should be studied by careful examination of patients, is embodied in today's more elaborate clinical investigations, as we shall see in Chapter 6.

But that is not enough for the vivisectors. Symptoms or diseases are artificially induced in animals in the hope that the resulting 'animal model' of human illness closely resembles the situation in patients. Yet how can such a crude approach be compared with the naturally or spontaneously arising illness in patients, particularly when the complex nature of disease – often involving emotional, genetic and environmental factors – is taken into account? The best we can hope for is that vivisection does not confuse or delay clinical studies with patients actually suffering from the disease.

Not surprisingly many doctors have questioned the basic fallacy of reproducing human disease in laboratory animals. In 1937 Sir Lionel Whitby, later to become President of the British Medical Association, wrote in the *Practitioner*: '. . . it is almost impossible, in an animal experiment, to reproduce a lesion or a disease at all comparable-to such as is found in the human subject.'[13]

Over 40 years late¬, despite the enormous increase in animal experiments, Professor Colin Dollery, a highly respected clinical scientist at the Royal Postgraduate Medical School in London, still had to acknowledge, '. . . for the great majority of disease entities, the animal models either do not exist or are really very poor.'[14] Nevertheless such artificially induced 'animal models' have proved distressingly popular. Animal models of anxiety, arthritis, atherosclerosis, cancer, cataract, depression, diabetes, epilepsy, heart attacks, high blood pressure, muscular dystrophy, obesity, stroke, syphilis, ulcers and many, many more have all been attempted.[15] And inevitably results often prove confusing.

In physiology '. . . subtle but important differences', writes the *Lancet*,[16] make it '. . . notoriously dangerous to apply experimental results from animals to the treatment of human beings.' Animal models of senile dementia have proved highly dubious and one scientist, examining 140 brains from 47 different species, found no senile lesions like those in people.[17] Even in normal ageing the situation in the animal brain appears to be very different to that in people. In the study of blood formation and its control, results from animal experiments, observes the *Lancet*, have '. . . rarely been transferable to human therapeutics.'[18] This is particularly the case with rodents whose reactions are sometimes 'strikingly different' from those in man.[19] Animal models of bronchitis have been produced by forcing rats to breathe the smoke of 25 cigarettes a day for 14 days,[20] but the human respiratory system is totally different to the rat's and, according to the scientific journal *Drug Metabolism Reviews:*

> 'The rat is not considered an appropriate model because its version of chronic respiratory disease that involves bronchitis displays excessive inflammation and involvement of the pulmonary paraenchyma.'[21]

Obesity has been artificially induced in mice and rats by injecting monosodium glutamate during the first week of life, by feeding a diet containing 60 per cent lard or by injecting a chemical compound of gold. Not surprisingly there were marked differences between the various animal models but the researchers also found that '. . . none of the models used are appropriate' to the human condition.[22] Tuberculosis in human patients and the artificially induced disease in animals are '. . . distinctly different' notes the *Lancet,*[23] though they are produced by the same organism. The disease is far more complex in people '. . . with a clinical and pathological picture very different from that in the guinea pig.'

Animal models of atherosclerosis have used birds, dogs, rats, pigs, rabbits and monkeys but species differences have been seen in all cases.[24] In people atherosclerosis takes many years to develop and results from fatty deposits in the arteries, eventually leading to angina, heart attacks, strokes and muscle and kidney problems. By feeding rabbits an unnatural, high-cholesterol diet, their arteries quickly become blocked but the lesions are quite different to those found in people, both in their content and distribution.[25] In the recently published *Animal Models in Cardiovascular Research,* Dr David Gross explains how, 'It is rare for lesions in rabbits to develop fibrosis, haemorrhage, ulceration and/or thrombosis, all characteristics of lesions in humans.'[25] In spite of these differences rabbits remain the most widely used species in atherosclerosis research because, as Gross points out, they are '. . . easy to feed, care for and handle. They are readily available and inexpensive.'

In animal models of stroke, the condition is artificially induced by constricting blood vessels with spring clips or ligatures or even by injecting tiny objects to simulate a blood clot. This bears little relation to the naturally arising disease in people, which takes a long time to develop and is caused by high blood pressure and atherosclerosis, where the blood vessels to the brain eventually become blocked. There are also fundamental differences in the anatomy of blood vessels in the brain that make animals a poor model for the human condition and far less prone to stroke. Discussing the role of animal experiments at a conference on Cerebral Vascular Diseases, Dr Whisnant, Consultant Neurologist at the Mayo Clinic in Rochester, states that,

'For the most part, these studies have tended to lag behind clinical and pathologic-anatomic investigation and too frequently have served as confirmatory work after clinical impressions have been virtually accepted.'
'It is obvious at the outset, that investigations with laboratory animals cannot be directly related to human disease. No experimental animal has an entirely comparable cerebro-vascular supply to that of man and most, if not all, have a considerably greater cerebral circulatory reserve. For example, we have used dogs for our studies and yet we have learned that both vertebral arteries and both common carotid arteries and all their terminal branches can be ligated in the neck simultaneously and still 50 per cent of the dogs survive and recover near normal function.'[26]

For decades scientists have induced ulcers in their animal victims. Sir Heneage Ogilvie, consulting surgeon to Guy's Hospital in London has strongly condemned the practice, arguing that peptic ulcers produced experimentally bear 'little resemblance' to those seen in patients.[27] Indeed, all Ogilvie's contributions were based on careful clinical observation of patients:

'I am particularly concerned not with the wickedness but with the folly of experiments on animals as a means of giving an answer to this particular question . . . To apply the results of experiments on dogs to the aetiology and treatment of peptic ulceration in man is as scientific as to base a course of postnatal lectures to mothers on a study of the maternal habits of the female kangaroo.'[28]

In 1918 Japanese scientists produced cancer on a rabbit's ear by painting it with tar, and a new era in cancer research had begun.[29] It was all too easy to induce cancer in laboratory animals, but for future generations of cancer patients, the discovery was to prove disastrous. Not only did it succeed in diverting attention from the causes of *human* cancer, but it fostered the belief that animal models of the disease would soon

Animal Experiments in Cancer Research

'It so happens that the whole of our knowledge of the structure, symptoms, diagnosis and treatment of the neoplasias [cancers] of man comes from those who approach the subject by direct clinical methods. To this extensive knowledge the contribution of laboratory experimentalists is practically nil.'

(Hastings Gilford, surgeon, *Lancet*, 1933)

'Research obtained with such artificial conditions might be expected to have little relation to human disease, for the genetic uniformity of the susceptible mouse families is never paralleled in man.'

(*Lancet*, 1951)

'I wonder how he interprets the fact that animal tumours show a remarkable tendency to spontaneous regression, so much so that in the opinion of leading cancer experts no safe conclusions can be drawn from therapeutical experiments on such animals. In man this tendency appears to be much less pronounced.'

(Dr F. M. Lehmann, *British Medical Journal*, 1952)

'We confess disappointment with the practical issues of experimental research in cancer. It has told us much about malignant tumours in the lower animals but this, if applied to man, does not tally with experience.'

(*Medical Officer*, 1952)

lead to a cure. Eventually it became clear that artificially induced cancers in animals were quite different to the spontaneous tumours that arise in patients. In 1951, *Medical Review* sadly observed:

'As the years pass, cancer seems to be on the increase. The search for the cause has up till now met with a very poor result, largely owing to the fact that cancer research has been and is being conducted on laboratory animals ... We believe that, until research switches over to the clinician and leaves the laboratory investigator to grieve over his failures, no real progress will be made.'[30]

But the warnings went unheeded and today the disease shows no sign of decline.[31] Animal tests mean that useful drugs might be missed whilst those curing the disease in mice could well prove useless in hospital trials. Professor Haddow, once Director of the Chester Beatty Institute at the Royal Cancer

---

Animal Experiments in Cancer Research

'... although lung tumours have been described in many species, there is no laboratory animal which spontaneously develops tumours comparable to the ordinary squamous or anaplastic carcinoma of the bronchus of man ...

'The application of these results to the study of human lung cancer is problematical. In the first place, pulmonary tumour of mice is histogenically different from bronchial carcinoma, and tumours which more closely resemble the human type have been produced only by methods which do not appear likely to have any direct counterpart in naturally occurring carcinogenesis.'

(Richard Doll,
British Medical Journal, 1953)

'Since no animal tumour is closely related to a cancer in human beings an agent which is active in the laboratory may well prove useless clinically.'

(*Lancet*, 1972)

'Indeed, while conflicting animal tests have often delayed and hampered advances in the war on cancer, they have never produced a single substantial advance either in the prevention or treatment of human cancer.'

(Dr Irwin Bross, Director of Biostatistics,
Roswell Park Memorial Institute for
Cancer Research, 1981)

'Indeed even while these [clinical] studies were starting warning voices were suggesting that data from research on animals could not be used to develop a treatment for human tumours.'

(*British Medical Journal*, 1982)

Hospital, points out that the beneficial effects of urethane in treating leukaemia were detected solely by clinical observation of patients.

> 'The various leukaemias in the mouse and rat were relatively refractory to the influence of urethane, and the remarkable effect in the human might have eluded discovery if attention had been directed to the animal alone.'[32]

Over a 25-year period, the United States National Cancer Institute screened 40,000 species of plants for antitumour activity and, as a result, several proved sufficiently safe and effective on the basis of animal tests to be included in human trials. Unfortunately all of these were either ineffective in treating human cancer or too toxic to consider for general use. Thus, in 25 years of this extensive programme, not a single antitumour agent safe and effective enough for use by patients has yet emerged.[33]

More enlightened researchers are now suggesting the use of human cancer cells and tissues – taken during therapeutic operations or after death – as a means not only of testing drugs but for understanding the nature of the disease itself. Results can then be directly matched with clinical and epidemiological findings and the misleading results from animal tests avoided. Already several substances rejected by animal experiments have proved highly active against human tumours.[34]

Human beings and animals also react differently to injury and shock. Clinical studies with patients show that people with burns should be kept warmer than usual to overcome the effects of shock. But such treatment is very hazardous to a burned rat[35] – it has entirely the opposite effect! Professor Stoner, until recently Director of Manchester's Trauma Unit, has described how dogs were the favourite animal for studying the traumatic effects of injury during the 1950s and 1960s. He explains how a scientist called Wiggers devised a method for producing haemorrhage in the dog that would be fatal. But according to Stoner, the 'model' does not bear much relationship to conditions seen clinically:

> 'This view was completely substantiated by Shoemaker, who, when he compared the cardiovascular patterns seen in patients with that seen in the dog in the Wiggers' model, found that there was no similarity at all. Nevertheless, this model has been studied by an enormous number of

physiologists. This is perhaps a reflection that nowadays there are many physiologists who have no contact with "real-life" medicine. Work of that sort is all right if you accept the fact that it has little relevance, and I think that the research that has been carried out on it has contributed very little to our understanding of the responses to injury.'[35]

Species differences have also been found in animal models of septic shock, caused by infection rather than injury. In laboratory animals septic shock is induced by injecting bacteria, and a chemical called naloxone has been suggested to alleviate the symptoms. The treatment works with dogs but not with baboons and indeed reports reveal striking differences in species response to naloxone therapy.[36] Once again the only way to discover if naloxone helps patients is to conduct careful clinical trials. Naloxone, it turns out, is no better than a placebo in ameliorating reduced blood pressure or improving survival in patients with septic shock.[37]

Whilst scientists try hard to produce convincing models of human disease, the real answer is always the same – the proper study of mankind is man.

Since 1876, when the Cruelty to Animals Act was passed, the use of animals to practise surgical skills has been prohibited, although the Home Office has licensed experiments to develop new techniques such as transplants. Human beings and animals are physiologically and anatomically different so it is indeed fortunate that British surgeons, recognized as amongst the best in the world, have not practised on animals. In the UK surgical skill is obtained by work with human bodies in the dissecting rooms, then by observation of senior surgeons during actual operations and, finally, by taking over under the close supervision of an experienced surgeon.

Until the mid nineteenth century advances in surgery were hampered by the fear of pain and by postoperative infection. The most valued surgeons were therefore the quickest! But with the discovery of anaesthetics, together with hygienic conditions, surgery advanced rapidly.[38] As clinical experience increased, techniques improved and operations, hitherto unthinkable, became possible. The battlefield also proved invaluable in developing new techniques and during the Second World War, surgery for wounds of the chest and heart became a relatively common procedure with the result that many of the fundamental skills of heart surgery were developed.[39]

One of the most brilliant surgeons of the late nineteenth century was Lawson Tait, to whom we owe so many of our present-day surgical techniques.[40] In 1868, when still only 23, he performed his first ovariotomy and by 1872 his name had gone into the history books with 'Tait's Operation' – the removal of the uterine appendages for chronic ovaritis; in 1877 he began to remove diseased Fallopian tubes and in 1878, still only 34, he performed the first successful cholecystectomy (a gall bladder operation) in Europe. In 1880 he was the first to sucessfully remove the appendix for the relief of appendicitis, a relatively common operation today. During the same year he challenged Lister's method of antisepsis using carbolic spray because of its damaging effects on the tissues, and became one of the first to promote today's aseptic surgery, based on absolute cleanliness. Indeed, Tait was renowned for his insistence on strict hygiene.

In 1883 he performed the first successful operation on a case of ruptured ectopic pregnancy (where the fertilized ovum grows within the Fallopian tube). The prognosis for untreated cases had been grim – of 149 cases only four had survived. Within the next five years Tait handled 40 such operations with only one death. By 1884 Tait had published a paper on his first 1,000 abdominal operations, at a time when laparotomy was still a rarity. Two years later he published a remarkable paper recording 139 consecutive ovariotomies without a single death – a surgical marvel for the period. He also worked for the admission of women into the medical profession on equal terms with men and his Presidential Address to the British Gynaecological Society in 1887 made a forceful plea for this equality of status.

These are but some of Tait's achievements and there is no doubt that in the years ahead, many thousands were to owe their lives to his innovative genius. Tait then was an expert surgeon but he did not owe his skill to animal experiments. On the contrary, he was a fierce critic of vivisection,[41] which he considered 'useless and misleading.' He believed that '. . . in the interests of true science its employment should be stopped, so that the energy and skill of scientific investigators should be directed into better and safer channels.' And further:

> 'The fact is that the diseases of animals are so different from those of men, wounds in animals act so differently from those of humanity, that the conclusions of vivisection are absolutely

worthless. They have done far more harm than good in surgery. In fact, the late Sir William Fergusson, Sergeant-Surgeon to the Queen, declared in his evidence that vivisection had done nothing at all for surgery, and I think his authority on such a subject almost beyond appeal.'[42]

Tait not only denounced vivisection in the public press but before his own colleagues. On April 20, 1882, he read a lengthy and detailed paper before the Birmingham Philosophical Society 'On the Uselessness of Vivisection upon Animals as a Method of Scientific Inquiry.' Referring to ovariotomy he writes:

'Disregarding all the conclusions of experiment, Baker Brown showed us how to bring our mortality of ovariotomy down to 10 per cent, and again in 1876, Keith proved that it might be still further reduced. The methods of this reduction were such as only clinical experience on human patients could indicate . . .'
'As soon as Keith's results were established, abdominal surgery advanced so rapidly that now, only six years after, there is not a single organ in the abdomen that has not had numerous operations performed upon it successfully. I have had, as is well known, some share in this advance, and I say, without hesitation, that I have been led astray again and again by the published results of experiments on animals, and I have had to discard them entirely.'[41]

But Tait was not alone. Other surgeons too were concerned that advances must be based on clinical experience rather than animal experiments.[43] Royal surgeon Sir Frederick Treves, writing in the *British Medical Journal* of 1898, issued a salutory warning:

'Many years ago I carried out on the Continent sundry operations upon the intestines of dogs, but such are the differences between the human and the canine bowel, that when I came to operate on man I found I was much hampered by my new experience, that I had everything to unlearn, and that my experiments had done little but unfit me to deal with the human intestine.'[44]

Tait's conclusion, that vivisection is not only useless but actually harmful, cannot be dismissed a hundred years later as irrelevant. His views were expressed at a time when surgical techniques were developing rapidly and apply equally in more modern times when transplants and other surgical feats are being contemplated. The crucial point is the underlying biological differences that make such experiments hazardous. The same warnings were being issued in the 1950s, this time by the *Lancet*.

> 'The gastro-intestinal tract in man is unfortunately very different from that of animals, and the results of a new operation for gastric disease cannot be predicted from operations on dogs.'[45]

More recently, thousands of animals have been used to develop transplant techniques but it is revealing that the first *human* operations are usually disastrous. Only after considerable clinical experience do techniques become more successful. Transplant patients also suffer the additional problem of rejection so they are given powerful drugs to suppress the body's natural defences. As a result a new medical problem is created and patients become highly susceptible to infections and to cancer. Transplant recipients are 100 times more likely to develop cancer than the rest of the population.[46]

In 1967 Christian Barnard performed the first 'successful' heart transplant – the patient surviving just 18 days.[47] Enthusiastic surgeons all over the world rushed to perform the new operation but, according to the *Lancet*,[48] results were 'largely disastrous.' One of the best-known heart transplant centres is at Stanford University in California. Over a nine-year period 400 operations were carried out on dogs yet the first human patients both died because of complications that had not arisen during preliminary experiments.[49] By 1980, 65 per cent of transplant recipients at Stanford were still alive after one year with the improvement almost entirely the result of increased skill with existing antirejection drugs and in the careful choice of patients for operation.[48]

The same is true of lung transplants: of the first 39 patients only two survived beyond two months.[50] This prompted operations for combined heart and lung transplants but, once again, the early experience was disastrous – the first three patients dying at 14 hours, 8 days and 23 days respectively after the operation.[50] In 1986 Stanford reported 28 heart and lung transplants carried out between March 1981 and August 1985.[51] Eight patients died during or immediately after the operation. In another ten a respiratory disorder called obliterative bronchiolitis (OB) developed after surgery from which four patients died and three were left 'functionally limited' by breathlessness. The surgeons note that '. . . extensive experience with animal models in this and other institutions had not indicated a serious hazard from airway disease, so the emergence of post-transplant OB as the most important complication was unexpected.' The remaining ten patients had no transplant complications although they all suffered kidney damage caused by cyclosporin, the drug used to suppress rejection. After three years just four patients remained free of complications. The doctors attributed their 'success' to the availability of cyclosporin, a continuing heart transplant programme and the development of an animal model!

In 1966 London's Westminster Hospital had performed 20 kidney transplant operations but only three patients survived beyond 66 days.[52] Yet the surgeons, led by Roy Calne, now at Cambridge, previously reported 'encouraging results' in experiments with dogs.[53] Nevertheless, as clinical experience developed, the situation improved although by 1980 44 per cent of transplanted kidneys still failed after one year. The recent introduction of a new drug, cyclosporin A, has made rejection less of a problem but ironically one of the drug's most serious side-effects is kidney damage. In 1984 the Stanford Heart Transplant Group reported serious kidney problems in 17 of 32 heart transplant patients treated with cyclosporin for more than a year, two of these patients ultimately requiring dialysis because their kidneys had failed.[54] Although cyclosporin-induced kidney damage is a real danger for patients, this side-effect was not seen in laboratory animals at therapeutic doses, except in a certain unusual strain of rat called Kyoto, specially bred to have reduced blood pressure.[55] But recent comparisons with cyclosporin G, a possible alternative to cyclosporin A, have led Calne and his colleagues to suggest that,

'. . . there are both species and strain differences in the absorption, metabolism, and excretion of these two cyclosporins and these differences influence nephrotoxicity [kidney damage].'[56]

One reason animal experiments produce the wrong answers is because of tissue differences between humans and animals. John Fabre of Oxford's Nuffield Department of Surgery, describes how positive results from animal experiments in the 1960s suggested that there might be important advances in transplantation and thereby prompted a large amount of further research into heart and kidney transplants in rats.[57] But tissue differences between humans and rats meant that animal experiments once again proved misleading:

> 'The many encouraging results raised hopes that a major advance in clinical immunosuppression for transplantation was in the offing, but these hopes have now faded and nothing of the great mass of work has been translated into clinical practice.'[57]

Not content with the transplant extravaganza, experimenters at the University of Utah have developed an artificial heart as a replacement for the patient's own ailing heart. The Jarvick 7 artificial heart, named after its inventor Robert Jarvick, was tested on calves, with one animal surviving for almost nine months after the operation.[58] These experiments led to the first human trial and in 1982 the stage was set when Barney Clark, a 61-year old dentist with heart disease volunteered to be the first human guinea pig. He survived a miserable 112 days. As he lay in his hospital bed, Dr Clark was tethered by 6 ft (1.83m) tubes to a huge external air compressor that powered his artificial heart. The machine, weighing 375 lbs (170 kg) was mounted on a trolley and had to be pushed around whenever Dr Clark was moved. But he never regained any mobility, nor the strength to push the trolley himself, and was plagued by a series of health problems from continual nose bleeds to kidney collapse. The kidney crisis that finally brought his death was the seventh major mishap during his hospital stay.[59]

But why stop at artificial hearts? Dr Leonard Baily at the Loma Linda Medical Centre in California hit the headlines in 1984 when he transplanted a baboon's heart into a two-week-old baby girl with heart disease. Baby Fae, as she came to be known, died 21 days later. At the time of the operation, it was reported that Baily had virtually no experience of human heart transplants but had performed about 160 cross-species transplants over the previous seven years, mainly in sheep and goats.[60] Apart from one transplant between two closely related sub-species of goat, none of the animals survived longer than six months – hardly a glowing testimony. And, according to Dr Martin Ruff, an immunologist at University College, London, rejection of the baboon heart was inevitable since there are no antigens in common between baboons and humans.[60] Antigens govern whether an organ is rejected and the absence of common antigens would thereby ensure rejection. Subsequent reports suggested that Baily had been determined to go ahead with the human experiment because no attempt had been made to find a human heart for Baby Fae.[61]

And then there is American neurosurgeon Dr Robert White whose macabre monkey-head transplants have even been condemned by the British and American Medical Associations.[62] White's experiments produced animals paralysed from the neck down because he was unable to rejoin the spinal cord . . .

If animal models of disease and injury often prove confusing, it is the safety testing of new drugs that so vividly illustrates the hazards of such work. Although a familiar defence of animal tests is that they actually protect the public from dangerous drugs, is this really the case?

Adverse effects of drugs are known to be grossly underreported yet during 1977 Government figures show that 120,366 patients were discharged from or died in hospitals in the UK after suffering the side-effects of medicinal products.[63] In general practice as many as 40 per cent of patients may experience harmful effects as a result of their treatment[63] and the medical textbook *Iatrogenic Diseases* observes that

> 'It is clear that adverse reactions are becoming an increasingly important problem in drug treatment. In Britain, perhaps 5 per cent of general hospital beds are occupied by patients suffering from their treatment. In the United States, an estimate has been given of one in seven hospital beds taken up by patients under treatment for adverse reactions caused by drugs.'[64]

So animal experiments cannot be doing a very good job. In fact, reliance on animal tests can prove dangerously misleading, actually *adding* to the burden of iatrogenic disease. To begin with animals simply do not have the potential to predict some of the most common or life-threatening side-effects.[65] For instance animals cannot tell us if they are suffering from nausea, headache, dizziness, amnesia, depression and other psychological disturbances and, although these are examples of relatively minor effects, they can still prove very troublesome for patients.

Mice, rats and rabbits – common laboratory animals – are physiologically unable to vomit yet this is one of the commonest side-effects: dogs are used instead to predict this adverse effect. Allergic reactions (often fatal as in the case of the withdrawn arthritis drug, Zomax[66]), some blood disorders, skin lesions and many central nervous system effects are examples of far more serious hazards that, once again, cannot be demonstrated in laboratory animals.

Sometimes genetic factors are at work that make individual patients more susceptible and the animal models of orthodox toxicity testing, writes a former Director of Wellcome Research Laboratories, '. . . give no basis for such subtle predictions.'[67] One of the world's best known toxicologists, Professor Gerhardt Zbinden from Zurich's Institute of Toxicology, states: 'Most adverse reactions that occur in man cannot be demonstrated, anticipated or avoided by the routine subacute and chronic toxicity experiment.'[68] Zbinden has shown that only three of the 45 most common drug side-effects might be predicted using animals.[69] Of the remaining 42 adverse effects, '. . . only in exceptional cases can they be predicted from routine toxicologic tests.'[69] So reliance on negative results from animal experiments as a measure of human safety is a serious mistake.

Even when we exclude the effects that cannot be demonstrated in animals, toxicity tests in various species still prove misleading. In 1962 the harmful effects of six different drugs, reported during the treatment of patients, were compared with those originally seen in rats and dogs.[70] Of 78 side-effects seen in the patients only 26 (one in three) also occurred in both rats and dogs, whilst 42 were seen only in the patients. So, in most cases, predictions based on animal experiments proved incorrect. Even then the comparisons were confined to those effects animal tests have the *potential* to predict – it omitted altogether effects such as nausea, headache and psychological disturbances because they can only be communicated by people. For one of the drugs 14 such symptoms were recorded.

Another study, published in 1978, revealed that *at best* only one out of every four adverse effects predicted by animal tests, actually occurred in patients.[71] Even then it is not possible to tell which predictions are correct until after clinical trials with volunteers. Furthermore the report confirmed that many common side-effects seen in patients cannot be predicted by animal tests at all: for instance, nausea, headache, dizziness, dry mouth, sweating, cramps, and in several cases, skin lesions and reduced blood pressure.

Not surprisingly then, many animal-tested drugs have been withdrawn or severely restricted when unexpected and even fatal reactions have occurred in patients. The drug Eraldin, marketed by ICI in the 1970s for the treatment of heart conditions, was found to cause serious eye damage, including blindness, and there were 23 deaths.[72] Eraldin was thoroughly tested and yet animal experiments gave no hint of the tragedy to come.[73] Even after the drug was withdrawn in 1976 no one could reproduce the harmful effects in laboratory animals.[67]

The antibiotic chloramphenicol, passed safe after animal experiments, was later discovered to cause aplastic anaemia, an often fatal blood disease. The *British Medical Journal* records how the drug was 'thoroughly tested' on animals, producing nothing worse than transient anaemia in dogs when given the drug for long periods by injection and nothing at all when given orally.[74] Yet 42 patients died in Britain alone [72] between 1964 and 1980 and chloramphenicol is now severely restricted to very serious infections.

As we have seen the arthritis drug Opren was withdrawn in 1982 after 3,500 reports of side-effects including 61 deaths,[3] mainly through liver damage in the elderly. According to an investigation by Granada TV's 'World in Action' programme, Eli Lilly insisted that they had no reason to think Opren would cause any particular problem for the elderly before they launched the drug.[167] Prolonged tests in rhesus monkeys (the species usually considered closest to us), in which the animals received up to seven times the maximum tolerated human dose for a year, revealed no evidence of toxicity.[75] Nor apparently had animal tests given any warning of the photosensitive skin reactions that were to bedevil patients during the drug's brief 22-month history.[76]

Several other arthritis drugs, including Alclofenac, Ibufenac, Flosint, Osmosin and Zomax, have also been withdrawn after unexpected, often fatal side-effects: Ibufenac lasted just two years, being withdrawn in 1968 after 12 deaths, mainly through liver damage, an effect animal tests again failed to predict.[77] And Flosint, marketed by Farmitalia Carlo Erba, survived a mere 12 months during which Britain's Committee on Safety of Medicines received 217 adverse reaction reports. There were eight deaths.[78] Prolonged tests in rats had shown 'excellent tolerability' and Flosint '. . . was also well tolerated in the dog and monkey in medium-term toxicity tests.' Furthermore the tests showed '. . . no signs of toxicity in both these species in long term tests.'[78]

In 1982 Astra introduced their new antidepressant drug

Zelmid but a year later the Committee on Safety of Medicines had received over 300 reports of adverse reactions.[63] Of these, 60 were serious, including convulsions, liver damage, neuropathies and eight cases of the Guillain-Barré syndrome that had never previously arisen as a drug side-effect. There were seven deaths. In September 1983 Zelmid was withdrawn yet the original tests in rats and dogs had shown no evidence of toxicity at five times the human dose.[80]

In the 1960s Japan suffered an epidemic of drug-induced disease associated with clioquinol, the main ingredient of Ciba Geigy's antidiarrhoea drugs Enterovioform and Mexaform. At least 10,000 people, and perhaps as many as 30,000, were victims of SMON (subacute myelo-optic neuropathy), a totally new disease whose symptoms include numbness, weakness in the legs, paralysis and eye problems, including blindness.[81] The effects are produced by nerve damage yet animal experiments carried out by the company and reported in the *Lancet*[82] revealed 'no evidence that clioquinol is neurotoxic', tests being carried out on rats, cats, beagles and rabbits. In 1970 Japan's Ministry of Health and Welfare banned the drug,[83] and later on in 1982, Ciba Geigy decided to remove Enterovioform and Mexaform from the world market.

Often, despite unexpected and harmful effects in patients, potentially valuable drugs are allowed to remain in use, the Committee on Safety of Medicines (CSM) alerting doctors with special warnings. Oral contraceptives were first used in Britain during 1963 but, as the years passed, unforeseen dangers began to emerge. Careful observation of women taking the pill revealed an increased risk of blood clots, causing heart attacks, lung disorders and strokes. Between 1964 and 1980 404 deaths were reported in Britain alone[72] and subsequently the pill's estrogen content was reduced. Not only had animal tests failed to identify the hazard but, in some species, oral contraceptives produced entirely the opposite effect, making it more difficult for the blood to clot! Professor Briggs of Deakin University in Australia states:

> 'Many experimental toxicity studies have been conducted on contraceptive estrogens, alone or in combination with progesterones. At multiples of the human dose no adverse effect on blood clotting was found in mice, rats, dogs, or non-human primates. Indeed, far from accelerating blood coagulation, high doses of estrogens in rats and dogs prolonged clotting times. There is therefore no appropriate

animal model for the coagulation changes occurring in women using oral contraceptives.'[84]

In 1976 and again in 1979 doctors were alerted to the dangers of clindamycin, a broad spectrum antibiotic that caused an often fatal intestinal disease known as pseudomembraneous colitis.[63] The drug was marketed in the UK during 1968 and by 1980 36 deaths had been reported.[72] Yet rats and dogs, given clindamycin every day for up to a year, could tolerate 20 times the maximum permitted human dose.[85] Another example is Janssen's antifungal drug, ketoconazole (Nizoral). In 1985 the Committee on Safety of Medicines issued a special warning of serious liver damage, reporting 82 cases with five deaths.[86] No evidence of liver toxicity had been found in the original animal tests.[87]

Yet another case is the anaesthetic halothane. During 1986 the CSM obtained the co-operation of manufacturers in strengthening the warnings of liver toxicity[168]: the drug had caused 150 deaths [72] between 1964 and 1980, but no evidence of liver toxicity had come from the initial animal tests.[97]

You might be wondering why there are not *even more* drug disasters and a still higher toll of side-effects! There are two good reasons. One is that side-effects are very much under-reported because doctors have little incentive to do so. Reports of adverse reactions come from medical journals (0.5 per cent), death certificates (2.4 per cent), correspondence (7.5 per cent) and drug companies (13.6 per cent), but mainly through special yellow cards (76 per cent) that doctors are asked to complete should they discover any harmful effects.[89] Unfortunately the system reveals only 1-10 per cent of drug side-effects[88] and was shown to be inadequate as long ago as 1976 by the Eraldin disaster.

Eraldin was prescribed for over four years before doctors realized that it caused serious eye damage. Then, within weeks of the first published report, 200 more cases came flooding in. Ultimately more than 1,000 patients received compensation for damage caused by Eraldin.[164] Even worse, only about a dozen of the 3,500 deaths linked with isoprenaline aerosol inhalers were reported by doctors at the time.[89] And only about 11 per cent of fatal reactions associated with anti-inflammatory drugs phenylbutazone and oxyphenbutazone, were reported.[89]

So the figures given earlier for drug-induced deaths are most probably only a small fraction of the true figure. One estimate[90]

puts the annual number of drug-induced deaths in Britain at
10,000-15,000 – four to five times the officially reported figure
of over 2,000 and about twice the number killed in road
accidents.[79] The estimate needs to be taken seriously if a recent
study on just *one* category of drugs – the non-steroidal
anti-inflammatory agents – is any guide. These products are
generally used to treat arthritis and related conditions but,
according to an editorial in the medical journal *Gut,* they may be
associated with more than 4,000 deaths every year in the UK
from gastrointestinal complications![165] Interestingly, Britain's
pharmaceutical industry estimates that, in general, medicines
cause around four *hundred* fatalities a year.[166] Doctors fail to
report side-effects for many reasons including complacency,
ignorance, and perhaps guilt because their prescribed treatment
has caused harm.

Apart from doctors under-reporting adverse effects of drugs,
there is also the fact that clinical trials with healthy volunteers
and patients weed out the great majority of harmful and
ineffective drugs *before* they reach the market. Ciba Geigy
estimate[91] that of every 20 chemical compounds found safe and
therapeutically effective in animal tests, only one ever becomes a
prescription drug! This is hardly surprising because, as we have
seen, animal models of human disease are usually very poor
whilst most adverse effects cannot be predicted by animal tests.
Clinical trials are the first really valid test of a new drug's safety
and efficacy and they reject 95 per cent of the products passed
by animal experiments.

Such trials are rarely given much publicity, but one example is
ICI's drug Fenclozic acid, which they hoped to market as an
antiarthritis pill.[92] Tests with rats, mice, dogs and monkeys gave
no indication of liver damage but in clinical trials some patients
suffered liver toxicity. The drug was stopped and the patients
recovered within a few days. Not content with the results of
human trials, ICI returned to the laboratory and subjected even
more species to the drug. Rabbits, guinea pigs, ferrets, cats,
pigs, horses, neonatal rats and mice, together with a different
strain of rat, were all tested and still no evidence of liver damage
could be found.

Another case is the drug Mitoxantrone, synthesized in the
hope of providing effective anticancer treatment without
side-effects on the heart.[93] Tests on beagle dogs 'failed to
demonstrate cardiac failure' but in clinical trials several patients

suffered side-effects including heart failure. Fortunately they recovered. Another potential anticancer drug Azauracil was well-tolerated by dogs and monkeys with no signs of toxicity to the nervous system.[6] In trials at one twentieth the dose, almost all patients developed central nervous system effects including coma, lethargy, mental deterioration, twitching, muscle weakness and hallucinations.

Even so, clinical trials involve relatively small numbers of people, so many harmful effects appear only once a drug is marketed and widely used. Even then, when serious and unexpected effects are revealed, as we have seen, drugs may still remain in use. Doctors are expected to appreciate the hazards and prescribe accordingly. So, misleading animal tests may lead to withdrawal or, as so often happens, to hasty warnings and reports in the medical press about unexpected toxicity. For instance, the cough suppressant Zipeprol was found to cause severe neurological symptoms at high doses, including seizures and coma, yet animal tests had failed to reveal the hazards.[94] The cutaneous and kidney problems seen with captopril, used to treat high blood pressure, were not predicted by animal experiments.[95] Captopril has also been linked with agranulocytosis, a very serious blood disease, but here the original animal tests proved even more confusing. Bone marrow suppression was observed in beagles but not rats, mice or monkeys so no doubt the experimenters were reassured – falsely as it turned out.[96]

A major disaster occurred in the UK during the 1960s when at least 3,500 young asthmatic sufferers died following the use of isoprenaline aerosol inhalers.[89] Isoprenaline is a powerful asthma drug and deaths were reported in countries using a particularly concentrated form of aerosol that delivered 0.4 mg of drug per spray.[98] Animal tests had shown that large doses of isoprenaline increased the heart rate but not sufficiently to kill the animals: cats could tolerate 175 times the dose found dangerous to asthmatics.[99] The drug was obviously thought to be safe as it was available over the counter at chemists' shops. From 1968, following the epidemic of drug-induced deaths, it could only be obtained on prescription. Even after the event it proved difficult to reproduce the drug's harmful effects in animals. Researchers at New York's Food and Drug Research Laboratory experimented on animals in a continuing assessment of isoprenaline aerosols. In 1971 they reported that,

'Intensive toxicologic studies with rats, guinea pigs, dogs and monkeys at dosage levels far in excess of current commercial metered dose vials ... have not elicited similar adverse effects.[100]

Scientists at Belfast's Queen's University injected increasing doses of isoprenaline into dogs at five to ten minute intervals, up to 20 times the dangerous human dose and still the animals did not die.[101] But enough of any substance will eventually cause death and one dog died after receiving 50 times the danger dose. Others have shown that, although the animals did not die, a very high dose of isoprenaline causes heart lesions in rats and hamsters but not dogs or pigs.[102]

As the evidence accumulates, it becomes difficult to see just how animal experiments protect the public from hazardous drugs. What then are the causes of these vast differences between the species? There are five basic stages in the action of a drug taken internally – absorption into the bloodstream, distribution to its site of action, mechanism of action, metabolism and excretion. And when it is remembered that people of different ages, sexes, health and genetic make-up may react quite differently, it is not surprising that other species often react very differently. Indeed, even a small change, repeated at each stage, can accumulate resulting in a major change of effect.

One of the most important factors is the speed and pattern of metabolism, or the way in which a drug is broken down by the body. Scientific reports show that *differences in drug metabolism between the species are the rule rather than the exception.*[103] The table below shows just how great these differences can be.

Toxic drug effects, not predicted by animal tests, may be seen in people if their metabolism is slower, resulting in longer exposure. The infamous anti-inflammatory drugs phenylbutazone and oxyphenbutazone are responsible for an estimated 10,000 deaths worldwide[105] and their use has finally been severely restricted. The chances of harmful effects occurring in people compared with laboratory animals are considerably increased because it takes much longer for patients to metabolize the drugs. In people it takes 72 hours to break down a dose of phenylbutazone but the corresponding times in rhesus monkeys, dogs, rats and rabbits are eight, six, six and three hours, respectively.[104] For oxyphenbutazone it takes 72 hours for people and only half an hour for dogs to metabolize the

Metabolism (in hours) of various drugs [104]

Drug	Human	Rhesus monkey	Dog	Mouse	Rat	Rabbit	Cat
Hexobarbitol	6		4.3	0.3	2.3	1	
Meperidine (Demerol)	5.5	1.2	0.9				
Phenylbutazone (Butazolidin)	72	8	6		6	3	
Ethyl biscoumacetate (Tromexan)	2		21			2	
Antipyrine	12	1.8	1.7				
Digitoxin	216		14		18		60
Digoxin	44		27		9		27

drug.[6] The time taken for Opren to be eliminated from the blood stream was much longer in elderly patients than in laboratory animals.[106]

Poisonous effects can also arise when patients form a toxic metabolite not seen in the original species used for tests. This often occurs because various species metabolize drugs in an entirely different way, not just at a different speed, and this represents a serious handicap for those relying on animal tests. A comparative study of 23 chemicals showed that in only four cases did rats and humans metabolize drugs in the same way.[107] One example is amphetamine, which is metabolized by the same route in people, dogs and mice (although faster in the mouse) but by a different pathway in the rat and by still another route in the guinea pig.[104] Tromexan (see the table above), an anticoagulant drug, is metabolized much faster by humans than by dogs, although by the same process. The rabbit metabolized the drug at the same speed as people but by an entirely different pathway.

To complicate matters, a drug or its metabolite may react with something being taken at the same time. In one case, serious side-effects and several deaths were caused when patients receiving antidepressant drugs ate certain cheeses. The drugs suppressed the metabolism of tyramine, a component of the cheese and turned a normally harmless ingredient into a really dangerous substance.[104]

Species differences are not only found during routine subacute and chronic toxicity tests, where the animals are dosed every day for weeks, months and even years, but in the more specialized areas such as the LD50, skin and eye irritancy tests, carcinogenicity experiments and tests for birth defects.

The LD50 Test

If there is one area, however, where many scientists and animal rights campaigners are united, it is in their condemnation of the notorious LD50 test. LD stands for 'lethal dose' and LD50 signifies the single dose necessary to kill 50 per cent of the animals used in the experiment. It has been widely used to test pesticides, cosmetics, drugs, weedkillers and household and industrial products. A common form of the test is by oral dosing using a tube inserted down the animal's throat. Other forms of dosing include injection, forced breathing of the vapour (LC50, lethal concentration 50 per cent), and application to the animal's skin. Mice, rats, rabbits, birds and fish are the usual species[108] whilst dogs, cats and monkeys have also been used occasionally.[169]

In the 'classical' or 'formal' LD50 several dose levels are administered to equal groups of male and female animals. At the lower doses few animals would die but at the highest doses nearly all the animals would be killed. The LD50 is then calculated from these results. An average of 60 animals is used, although this would only be for one species and one particular dosing method.[108] Fewer animals would be used for an approximate LD50.

The test is allowed to proceed for 14 days, assuming the animals have not already died. They are then killed and sometimes their tissues examined for signs of poisoning. Convulsions, tears, diarrhoea, discharge and bleeding from the eyes or mouth and 'unusual vocalization' are typical symptoms in the dying animals.[109] As the British Toxicology Society explained,

> '. . . there is pressure on the toxicologist to allow the study to continue, even when the animals are in distress since their premature killing may alter the end-point of the study, and so possibly affect the classification of the material being tested.'[110]

Enough of any substance, however harmless, will cause undesirable effects, and death may be caused by overpowering the animal's ability to cope with the sheer quantities given rather than by any particular poisonous action of the chemical. The kind of grotesque experiment in which animals are dosed with huge quantities of harmless chemicals thereby overloading one or more of the body's organs, and finally causing death, is totally unrelated to the human situation, yct numerous examples have been reported in the scientific literature. What makes the tests even more unreal is that, unlike humans, rats and mice cannot vomit to remove the chemical from their stomachs.

In 1980 the journal *Archives of Toxicology* described how tartaric acid, a preservative in wines and baked foods, had been subjected to LD50 tests.[111] Doses of 2.5-32.2 grams (per kilogram of the animal's body weight) caused death but not because of a true toxic effect: the huge dose of an acid had caused serious corrosion of the stomach lining. Put into human terms, a person would have to consume up to 5 lbs (2.25 kg) in one dose to cause death. And since tartaric acid is normally present only in minute amounts this would mean consuming up to 3,960 pints (2,250 litres) or 1¼ tons of food all at one sitting!

Another recent example was quoted in the *Journal of the American College of Toxicology*. Animals were made to consume huge amounts of cosmetics, particularly lipsticks and waxes. In one experiment rats were force-fed up to 25g/kg of several lipstick formulations, the human equivalent of 4 lbs (1.8 kg).[112] For one product '. . . one animal died of intestinal obstruction but no toxic effects were seen'. The next time sugar is added to your tea or coffee remember that according to experiments with rats it would take 5½ lbs (2.5 kg) all in one dose to kill a person.[113] Experiments like this have led scientists to advocate 'limit' tests, which means that if the animals have not died after a predetermined, reasonably high dose, then the tests would be stopped.

The initial purpose of the LD50, when introduced by Trevan in 1927, was to measure the strength of drugs like digitalis but, as scientists used their imagination and designed better (non-animal) methods, the LD50 became obsolete. Unfortunately it is so easy to perform that scientists started using it as a crude index of toxicity: if substance A had an LD50 higher than substance B, A was considered less toxic as more was needed to kill the animals. Crude indeed but its very simplicity meant that it soon became part of a check list, included in data sheets like other pieces of information such as chemical formula or solubility. The idea of a single numerical index of toxicity naturally appealed to government bureaucrats and the LD50 became cnshrined in official government requirements for a wide range of substances including drugs, pesticides and

Paraquat – Dermal LD50 in rabbits (Hazleton 1980)		
Animal number and sex	Day(s) of study	Observations

Individual clinical observations: Group 1 – 62.5 mg/kg
Paraquat dichloride formulation

Intact

5140M	1	Lethargy
	2	Loose stools
	2-14	Brown perianal staining
5141M	1	Large haemorrhage from penis, animal killed in extremis
5148F	2	Hunched posture
	2-13	Lethargy
5149F	2	Lethargy, hunched posture

Abraded

5156M	1	Lethargy
5157M	1	Lethargy
	7	Weight loss
5164F		No clinical changes observed
5165F	1-2	Lethargy, loss of hind limb cooordination
	1	Animal quivering, all hind limb muscles held in contracted position
	2	Large anal haemorrhage, animal killed in extremis

Individual clinical observations: Group 2 – 125 mg/kg
Paraquat dichloride formulation

Intact

5142M	1	Animal prostrate on cage floor and total loss of mobility – killed in extremis
5143M	1	Found dead
5150F	1	Lethargy
	2-3	Profuse perianal staining
	4-13	Brown perianal staining
	14	Cream coloured perianal staining
5151F	1	Animal killed in extremis

Abraded

5158M	1-2	Lethargy
	1	Reduced respiration rate
	2	Limb movements uncoordinated. Severe respiratory distress – animal killed in extremis
5159M	1	Animal killed in extremis

Animal number and sex	Day(s) of study	Observations
5166F	*1*	*Animal killed in extremis*
5167F	*1-2*	*Lethargy. Uncoordinated hind limb movements*
	2	*Animal prostrate and unable to maintain normal posture. Laboured breathing and reduced respiration rate. Killed in extremis*

Individual clinical observations: Group 3 – 250 mg/kg Paraquat dichloride formulation

Intact

5144M	*1*	*Lethargy. Impaired movement of hind limbs*
	2	*Found dead*
5145M	*1*	*Found dead*
5152F	*1*	*Lethargy*
	1	*Found dead*
5153F	*1*	*Found dead*

Abraded

5160M	*1*	*Found dead*
5161M	*1*	*Found dead*
5168F	*1*	*Animal killed in extremis*
5169F	*1*	*Animal killed in extremis*

Individual clinical observations: Group 4 – 500 mg/kg Paraquat dichloride formulation

Intact

5146M	*1*	*Lethargy. Total loss of mobility and prone on cage floor – killed in extremis*
5147M	*1*	*Animal killed in extremis*
5154F	*1*	*Found dead*
5155F	*1*	*Animal killed in extremis*

Abraded

5162M	*1*	*Animal killed in extremis*
5163M	*1*	*Found dead*
5170F	*1*	*Found dead*
5171F	*1*	*Found dead*

'. . . the LD50 is a crude measure of toxicity. There is really little scientific justification for the test because reproducibility is not good, it can even vary from day to day, and the results are dependent on the animal strain used.'

(Phillip Rogers, Managing Director, Hazleton Laboratories, 1977[121]) industrial products. According to one of Britain's largest contract laboratories, Huntingdon Research Centre, 'Approximately 90 per cent of LD50 tests which are performed by this Contract Research Centre, and probably by others also, are purely to obtain a value for various legislative needs.'[115] And the Director of Shell Toxicology Laboratory has stated, 'The legislative value of this test is far greater than is its scientific value.'[116]

There was of course a small, but by now familiar problem: results varied widely between species and even between different strains of the same species. The LD50 for digitoxin in rats is 670 times that in cats,[10] whilst for the antifungal substance antimycin A, the LD50 in chickens is 30 – 80 times greater than in pigeons and mallards.[117] The LD50 of thiourea in the wild Norway rat is 450 times greater than in the Hopkin's strain of rat.[118]

As long ago as 1948 Müller compared the lethal doses of various chemicals in animals and people and found frequent extreme variations (see the table opposite). Occasionally the results were roughly similar but usually quite different,[118] making it impossible to predict the human lethal dose, which can only be assessed by cases of accidental or intentional overdose. Nor could the test be used to predict symptoms or the outcome of overdose as experiments with paracetamol showed. The drug causes death in mice and hamsters by liver damage (LD50 250-400 mg/kg) but in rats the LD50 is considerably higher (1,000 mg/kg) and even then it is hardly possible to see liver damage.[111] Dr Roy Goulding, who established the first British National Poisons Information Service at Guy's Hospital in London, has stated:

> 'Whilst the data from animal studies . . . provide some basic information of the mechanism of toxicity and relative toxicity, it cannot be assumed that this information will be entirely relevant for man.
> 'Experience gained from a careful assessment of patients suffering from acute overdose of drugs is potentially much more useful than that obtained from animal tests.'[119]

In an account of how the National Poisons Centre at New Cross Hospital in London collates information and prepares advice on the prevention and management of drug overdose, the Director, Dr G. N. Volans, demonstrated that '. . . acute toxicity data [the LD50 is an acute toxicity test] from animal tests contributes

Comparison of the lethal doses found in animal experiments and in man (from Müller 1948)[118]

Sensitivity of man
compared to animal
experiment

Sensitivity of man
compared to animal
experiment

Stimulants
Pentylenetetrazol	1x
Caffeine	1x
Picrotoxin	1x
Strychnine	1x

Chemotherapeutics
Emetine	1x
Sufanilamide	2-4x
Quinine	6-8x
Arsphenamine	2-30x

Blood poisons
Aniline	1x
Potassium cyanide	1x
Hydrocyanic acid	1x
Potassium chlorate	5-7x

Metabolic poisons
Oxalic acid	10-20x
Salicylic acid	10-20x
Arsenic	3-40x
Phosphorus	10-60x

Antipyretics-Analgesics
Aspirin	1x
Aminopyrine	1 1/3-2x
Antipyrine	5-10x

Disinfectants
Potassium permanganate	1x
Thymol	ca.10x
Mercuric chloride	3-12x
Iodoform	4-100x

Hypnotics
Phenobarbital	1x
Tribromoethanol	2-3x
Propallylonal	3-4x
Cyclobarbital	1.5-5x
Carbromal	2-5x
Diallyl barbituric acid	3-6x
Chloral hydrate	10x
Barbital	3-15x
Sulfonmethane	6-18x

Local anaesthetics
Dibucaine HCl	2-5x
Tetracaine HCl	7-12x
Alypin	3-30x
Tropacocaine	20-70x
Cocaine	4-100x
Procaine	30-150x

Autonomic nervous system drugs
Physostigmine	1x
Epinephrine	10-15x
Atropine	600-1000x
Pilocarpine	500-2000x

very little of value to this work.'[170] According to their LD50s in animals, aspirin should be safer in overdose than another painkiller, ibuprofen. Yet, clinical experience does not accord with this since aspirin can kill people at doses that are not difficult to take, whilst the largest doses of ibuprofen recorded in over 14 years' clinical experience failed to produce serious toxicity.[170] Choice of the safest drug on the basis of its LD50 in animals could, therefore, be dangerously misleading.

Animal experiments cannot prevent accidental poisoning but child-resistant containers certainly can. Hospital admissions after accidental poisoning with analgesics fell dramatically after the introduction of child-resistant packaging in 1976.[120]

Neither can LD50s be used as a guide to select dose levels for the more prolonged animal tests because poisonous effects of repeated dosing cannot usually be predicted from a test such as the LD50, which uses a single dose. For instance, in rats the LD50 of the corticosteroid hormone, dexamethazone, is 120 mg/kg but on repeated administration, rats and dogs could not tolerate daily doses above a minute 0.07 mg/kg, approximately 1,700 times less than the LD50 value.[118] Consequently LD50s could also be dangerously misleading if used to predict doses suitable for repeated administration to human volunteers.

LD50s are not only influenced by species and strain but many other factors such as sex, age, degree of starvation, method of dosing, temperature, humidity and even bedding material! In mice allowed free access to food, the oral LD50 of the barbiturate sodium methohexiton is 354 mg/kg. When food is withheld for four to six hours the LD50 falls to 162 mg/kg whilst after 20 hours food deprivation the LD50 was only 66 mg/kg.[118] The number of animals per cage can also affect results. The LD50 of isoprenaline decreased from 800 mg/kg to 50 mg/kg as a result of isolating the animals for three months.[118] The LD50 can hardly be called a biological constant and results for the same chemical can vary from lab to lab by as much as 8-14 times, using the same species and the same method of dosing![125]

Over the past ten years scientists have become very critical of the test and in a major review, Professor Zbinden of Zurich's Institute of Toxicology concluded that:

> 'For the recognition of the symptomatology of acute poisoning in man, and for the determination of the human lethal dose, the LD50 in animals is of very little value.'[111]

Scientists and the LD50 Test

'... even if the LD50 could be measured exactly and reproducibly, the knowledge of its precise numerical value would barely be of practical importance, because an extrapolation from the experimental animals to man is hardly possible'

(D. Lorke,
Institute of Toxicology, Bayer AG,
Federal Republic of Germany, 1983[122])

'As an index of acute toxicity, this is valueless.'

(Dr Sharratt,
British Petroleum, 1977[123])

'... as far as new medicines are concerned, industry sees no general need for the LD50 value except in special circumstances, and some regulatory authorities are moving away from their former reliance on it.'

(Association of the British
Pharmaceutical Industry, 1984[124])

'For the selection of doses to be used in subacute and chronic toxicity experiments, the LD50 test does not provide consistent and reliable results.'

(G. Zbinden & M. Flury-Roversi,
Institute of Toxicology, Zurich, 1981[111])

The Draize test

Shampoos, pesticides, weedkillers, household detergents, even riot control gases, are just some of the products tested for irritancy in the eyes of conscious rabbits. The Draize test, introduced in 1944, is cruelly simple. The substance under test is sprayed or instilled into one eye of an albino rabbit. Generally no pain relief is given[126] and the test often proceeds for seven days,[127] during which the cornea, iris and conjunctivae are examined for signs of opacity, ulceration, haemorrhage, redness, swelling and discharge.

The very least one might expect from such a distressing

Scientists and the Draize Test

'There are important structural and biochemical differences between the human eye and that of animals, and extreme caution is required in extrapolating the results from animals to the likely condition in man. A typical example is the differing response of the eye of man and animals to repeated topical [surface] application of corticosteroids. Such a procedure is without effect on tension of the eye of many experimental mammals, but increases tension in the human eye.'

(B. Ballantyne & D. W. Swanston,
Porton Down, 1977[128])

'The predictive reliability of this technique has been questioned and its use of living animals has been criticized.'

(C. K. Muir, C. Flower & N. J. Van Abbe,
Leicester Polytechnic & Beecham
Products Research Department, 1983[131])

'. . . the rabbit eye is structurally and physiologically different from the human eye.'

(R. B. Kemp, R. W. J. Meredith, S. Gamble
& M. Frost, 1983
University College Wales, Ortho-
Cilag Pharmaceuticals Ltd and
Johnson and Johnson Ltd, 1983[132])

'. . . one of the important areas where it is difficult to extrapolate animal data to man is the eye and skin.'

(F. Coulston & D. M. Serrone,
Institute of Experimental Pathology &
Toxicology, New York, 1969[129])

'No single animal species has been found to model exactly for the human eye either in anatomical terms or in response to irritation.'

(D. W. Swanston, 1983[133])

procedure is that results safeguard the public. But the test has been condemned on scientific grounds because it produces unreliable results that often bear little relation to human responses. The rabbit is commonly used in eye irritancy tests because it is cheap, easy to handle and has a large eye for assessing results.[128] But there are major differences that make the rabbit eye a bad model for the human eye:[129]

1. Unlike people, the rabbit has a nictitating membrane, or third eyelid. It could be argued that this removes irritants from the eye or, on the other hand, serves as a trap for material beneath it.
2. The rabbit produces tears less effectively than people, so differences in the degree and duration of contact of an irritant can be expected. This may affect test results.
3. The acidity (pH) and buffering capacity of the aqueous humour in the eyes of human beings and rabbits are different. In humans, the pH is 7.1-7.3 whilst in rabbits it is 8.2. It is thought this difference might explain the particular susceptibility of the rabbit iris to chemical inflammation.
4. The thickness, tissue structure and biochemistry of the rabbit and human cornea are different. The thickness of the cornea in humans is 0.51 mm and 0.37 mm in rabbits.

So it comes as no surprise when the test proves misleading. In one case a liquid anionic surfactant formulation, with a long history as a basic ingredient of light duty dishwashing products, caused severe eye irritation in rabbits but extensive human experience after accidental exposure has shown it to be completely non-hazardous.[130] On the other hand when the sensory irritants CS and CR were tested in the eyes of human beings and rabbits, enormous differences were found with people 90 times more sensitive to CR and 18 times more sensitive to CS than rabbits.[133] According to scientists at Proctor and Gamble,

> 'It has not been possible for us to use the results of rabbit studies to predict accurately the actual irritation that might occur in humans after accidental exposure.'[130]

Animal tests can also be misleading in devising treatment for eye irritation. Clinical experience in treating human eye burns, caused by alkali, led to the preferred treatment of thorough rinsing followed by complete denudation of the cornea, this being successful if carried out within two hours. In rabbits, the same technique was unsuccessful, in fact denudation actually retarded recovery three-fold![130]

Another major problem is that results for the same chemical can vary widely from laboratory to laboratory and indeed within the same laboratory because of the subjective nature of assessing the test results.[127] What is classed as a severe eye irritant by one scientist may be dismissed as a mild irritant by another. This was the outcome of a study published in 1971, in which 25 laboratories tested 12 chemicals of already known irritancy. 'Extreme variation' between laboratories was found and the authors concluded that:

> 'The rabbit eye (and skin) procedures currently recommended by the Federal agencies ... should not be recommended as standard procedures in any new regulations. Without careful re-education these tests result in unreliable results.'[134]

Only when animal protection groups focussed public attention on the Draize test did attitudes start to change: now there are several test tube alternatives that could replace the Draize test immediately.[135]

Similar problems have been found during skin irritancy tests. Comparative trials by the UK's Huntingdon Research Centre with six species – mice, guinea pigs, minipigs, piglets, dogs and baboons – revealed '... considerable variability in irritancy response between the different species.'[171] The greatest differences were found with the more irritant substances. For instance, an antidandruff cream shampoo caused severe irritation in rabbits but only mild irritation in human volunteers. In baboons there was virtually no irritation at all.

Tests for Cancer

Carcinogenicity tests are supposed to identify substances likely to cause cancer in people and for economic reasons, the species most commonly used are rats and mice. Since the tests last around three years, they are extremely expensive and cannot possibly cope with the more than 40,000 largely untested chemicals currently in use. A recent study[136] compared the results of such tests carried out in mice and rats and found that 46 per cent of the substances carcinogenic in one species were found to be safe in the other! The conclusion is inevitable:

> 'It is painfully clear that carcinogenesis in the mouse cannot now be predicted from positive data obtained from the rat and vice versa.'

If it proves impossible to extrapolate results from rats to mice, how then can such tests give reliable information about effects in human beings, particularly in view of the massive doses used in animal tests, the lifetime of exposure and the differences in metabolism between animals and people?

Another study, published by Pfizer's David Salsburg, found that '. . . the lifetime feeding study in mice and rats appears to have less than a 50 per cent probability of finding known human carcinogens.' On the basis of probability theory, Salsburg writes, we would have been better off to toss a coin.[114]

Species differences in cancer tests

Chemical	Carcinogen	
	Yes	No
Arsenic	Humans	Mice, rats[7]
Benzene	Humans	Mice[7]
Iron oxide dusts	Humans	Hamsters, Mice guinea pigs[7]
2-naphthylamine	Dogs, monkeys	Rats, rabbits[137]
Pronethal	Mice	Rats, guinea pigs, dogs[138]

Thalidomide and Tests for Birth Defects

History's most infamous drug disaster left 10,000 children crippled and deformed. The culprit was thalidomide, marketed initially as a sedative by the German drug company Chemie Grünenthal in 1957, and by the Distillers Company in Britain a year later. Its ready clinical acceptance was no doubt based on an apparent lack of toxicity – animals could tolerate massive doses in routine tests without ill-effects.[63] Then came reports of peripheral neuritis that revealed thalidomide's toxic effects on the nervous system. This is a serious illness and may begin with a prickly feeling in the toes, followed by a sensation of numbness and cold. The numbness spreads and is eventually followed by severe muscular cramps, weakness of the limbs and a lack of coordination.

Even when the drug is stopped, the symptoms do not always disappear. This unexpected and serious side-effect had not, it seems, been predicted by the original animal tests but had the company now withdrawn the drug, a major disaster could have been avoided. Unfortunately the drug remained in use and a year after the initial reports of peripheral neuritis came the first indication that thalidomide damaged the human foetus. In a letter to the *Lancet* (December 16, 1961) Dr W. G. McBride, an obstetrician practising in Sydney, described severe malformations in babies whose mothers had taken thalidomide during pregnancy. 'Have any of your readers', he asked, 'seen similar abnormalities in babies delivered of women who have taken this drug during pregnancy?'

By June 1961, five months before his warning in the *Lancet*, McBride had seen three babies with unusual deformities and had strongly suspected thalidomide. To test his suspicions, he commenced experiments with guinea pigs and mice but no malformations were found and he started to have doubts that were to nag him for months.[139] Then, during late September came further malformed babies and McBride became certain that thalidomide was responsible. The animal tests must have been wrong. He wrote to the *Lancet* and the *Medical Journal of Australia* and alerted the medical profession. McBride's announcement led to more animal testing, this time by Dr Somers, a pharmacologist at the Distillers Company, but once again the first experiments proved unsuccessful.[63] Somers had chosen rats but no malformations were found and it was only when New Zealand white rabbits were used that thalidomide produced birth defects similar to those found in human babies. Somers had indeed been fortunate: he had hit upon one of the few species, other than human beings, in which thalidomide's terrible effects can be demonstrated.

It is true, thalidomide had not been specifically tested for birth defects before being marketed and, in order to prevent another disaster, the argument ran, new drugs must first be tested on pregnant animals (teratogenicity tests; teratogenic literally meaning monster forming).

At first this sounds plausible until we look more closely at the thalidomide tragedy. Even if the drug had been tested on pregnant rats, the animals so often used to look for foetal damage, no malformations would have been found,[6] as Somers was to discover later on. Thalidomide does not cause birth defects in rats or in many other species. The human tragedy would have occurred just the same. Commenting on this most infamous of drug disasters, the *Catalogue of Teratogenic Agents* (a scientific compilation of chemicals known to cause birth defects) stated that,

'... several ... principles were forcefully illustrated by observations made of the outbreak. The first point was that there existed extreme variability in species susceptibility to thalidomide.'[140]

By 1966 there were 14 separate publications describing the effects of thalidomide on pregnant mice. Nearly all reported negative findings or else a few defects that did not resemble the characteristic effects of the drug.[140] McBride's experiments with mice when attempting to confirm his suspicions about the drug, had also been negative. Writing in his book *Drugs as Teratogens*, J. L. Schardein observes:

'In approximately 10 strains of rats, 15 strains of mice, eleven breeds of rabbit, two breeds of dogs, three strains of hamsters, eight species of primates and in other such varied species as cats, armadillos, guinea pigs, swine and ferrets in which thalidomide has been tested teratogenic effects have been induced only occasionally.'[141]

Only in certain types of rabbit and primate could thalidomide's notorious effects be seen. So if thalidomide taught us anything it should have been that different species can react very differently and that women should only take drugs during pregnancy when absolutely essential. Thalidomide, prescribed as a sedative or for morning sickness, was hardly essential for life or health.

As the thalidomide tragedy recedes, scientists have become more familiar with teratogenicity tests and they are now widely believed to have little value. In pregnant animals, differences in the physiological structure, function and biochemistry of the placenta aggravate the usual differences in metabolism, excretion, distribution and absorption that exist between species and make reliable predictions impossible. For instance, of 800

chemicals identified as teratogens in laboratory animals, less
than 25 are known to harm people.[142].

Species differences in tests for birth defects

Drug	Teratogen	
	Yes	No
Aspirin	Rats, mice, monkeys, guinea pigs, cats, dogs	Humans[5]
Aminopterin	Humans	Monkeys[143]
Azathioprine	Rabbits	Rats[84]
Caffeine	Rats, mice	Rabbits[84]
Cortisone	Mice, rabbits	Rats[84]
Thalidomide	Humans	Rats, mice, hamsters[6]
Triamcinalone	Mice	Humans[140]

But the tests do serve a function. Dr Peter Lewis, a Consultant
Physician at the Hammersmith Hospital in London believes
teratogenicity tests are '... virtually useless scientifically.'
Nevertheless, he argues, they do provide

> '... some defence against public allegations of neglect of
> adequate drug testing. In other words, "something" is being
> done, although it is not the right thing.'[144]

And D. F. Hawkins, Professor of Obstetric Therapeutics at the
Institute of Obstetrics and Gynaecology and Consultant
Obstetrican and Gynaecologist at Hammersmith Hospital,
reveals the real motivation behind the tests:

> 'The great majority of perinatal toxicological studies seem to
> be intended to convey medico-legal protection to the
> pharmaceutical houses and political protection to the official
> regulatory bodies, rather than produce information that might
> be of value in human therapeutics.'[5]

The problem is highlighted by aspirin, a proven teratogen in
rats, mice, guinea pigs, cats, dogs and monkeys yet, despite
being widely used by pregnant women, has failed to produce any

kind of characteristic malformation.[63] The inescapable conclusion is that the tests are being performed not for scientific or medical purposes but for political and legal reasons – a cruel and illogical response to public concern arising from the thalidomide disaster. R. W. Smithells, Professor of Paediatrics and Child Health at the University of Leeds and a former member of the Committee on Safety of Medicines, aptly sums up the position:

> 'The extensive animal reproductive studies to which all new drugs are now subjected are more in the nature of a public relations exercise than a serious contribution to drug safety ... The illogicality of the situation is demonstrated by the continued use of well-established drugs which are known to be teratogenic in some mammalian species (e.g. aspirin, penicillin/streptomycin, cortisone). Conversely a new drug which comes through its animal reproductive studies with flying colours may nevertheless be teratogenic in man.'[145]

Whilst animal experiments often give a misleading impression of drug safety, they must also lead to the rejection of potentially valuable medicines on the basis of side-effects that would never occur in people. The finding that *at best* only one out of every four side-effects predicted by animals actually appears in patients, proves the point.[71] A good example is penicillin, which would in all probability have been discarded had it been tested on guinea pigs,[6] a species commonly used during infection research. Fleming's famous discovery, made in 1928, did not use animals, of course, but the humble culture dish. Fleming noticed that bacteria would not grow on a culture medium accidentally contaminated by a mould, *penicillium notatum*, which had floated into the laboratory through an open window. Evidently the mould produced something poisonous to the bacteria and Fleming called it penicillin. Even before Fleming's chance discovery, families kept the mould on damp cheeses to make a plaster for infected wounds. The mould was later shown to be a source of penicillin.[146]

Fleming was worried that penicillin might be de-activated by blood and his worst fears seemed to be confirmed when he injected a sample into rabbits. In fact the rabbit rapidly excreted penicillin in its urine but this was not realized at the time. The result so discouraged Fleming that he progressively lost interest and restricted penicillin for use in surface infections.[147] Later, Oxford scientists Florey and Chain resurrected the drug and found it cured deliberately infected mice. Florey later commented,

> '. . . mice were tried in the initial toxicity tests because of their small size, but what a lucky chance it was, for in this respect man is like the mouse and not the guinea pig. If we had used guinea pigs exclusively we should have said that penicillin was toxic, and we probably should not have proceeded to try to overcome the difficulties of producing the substance for trial in man.'[148]

But the good luck did not end there. In order to save a dangerously ill patient, Fleming wanted to inject penicillin into the spine but the results of such administration were unknown. Florey tried the experiment with a cat but there was not time to wait for the results if Fleming's patient was to have a chance. Fleming's patient received his injection and improved, but Florey's cat died . . .[147]

In fact, many useful drugs have been introduced into therapeutics without previous animal experiments. Examples include digitalis for heart conditions; quinine for malaria; ipecac for amoebic dysentery; and the early inhalation anaesthetics that, together with hygiene conditions, enabled surgery to advance rapidly. In a major review of the problems of species variation, American scientists Koppanyi and Avery state:

> 'Had these drugs first been tested in animal experiments for their safety, some of them might never have reached clinical trial.'[6]

So it is not surprising that animal experiments have often confused the results of clinical observation. The beneficial effects of digitalis as a treatment for heart conditions have been known for many years but its widespread use was delayed because animal experiments indicated a dangerous rise in blood pressure.[150] Eventually clinical observation of patients showed this to be incorrect.

Another example is iron sorbitol, used to treat iron deficiency anaemia. Intramuscular injections in rats and rabbits caused sarcomas at the site of injection and the implication for human therapeutics appeared serious. But 20 years after the initial observation in rats, clinical experience had revealed no real hazard to patients.[67] Salbutamol, used to treat asthma and bronchospasm, was found to cause tumours in rats (but not mice) and clinical trials were halted in the USA. But the drug had been available for several years in Britain and no evidence of cancer in human beings had been found.[151] Frusemide, a well tolerated diuretic in people, was subsequently found to cause severe liver damage in mice because of a metabolite not formed to any serious extent in people.[67] And 6-azauridine, tolerated for relatively long periods in the treatment of human cancer, was found to cause potentially lethal bone marrow depression in dogs after seven to ten days, even though much smaller doses were used.[67]

We have already seen how differences in metabolism between the species can lead to unforeseen toxic effects. But, according to Bernard Brodie of Bethesda's National Heart Institute, such differences also make it difficult to assess the efficacy of drugs and thereby '. . . create a serious obstacle to the development of new therapeutic agents.' He quotes Meperidine, a narcotic analgesic:

> 'Meperidine . . . is metabolized in man at the rate of about 17 per cent per hour . . . In order to study its fate and distribution, G. G. Burns of our laboratory decided to give a dog 20mg per kilogram, which in man would be a huge dose. He was prepared to use artificial respiration, since he fully expected that there would be dire effects including respiratory failure. The drug was infused over a period of 20 minutes, after which to his surprise the dog leaped from the table and walked away. The dog was relatively resistant to the effects of this narcotic because of the extraordinary rapid rate of biotransformation of Meperidine, 70 per cent to 90 per cent per hour, in this species . . . One wonders how many analgesics useful for man which showed only minimal effects in test animals are still sitting neglected on the dusty shelves of pharmaceutical houses only because they were metabolized too rapidly in the particular species in which they were tested.'[4]

In another example Brodie describes the anticoagulant drug ethyl biscoumacetate. Fortunately, he writes, the drug was studied in the rabbit, which metabolizes it at about the same speed as man. But the similarity in metabolic rate is sheer coincidence since the drug is metabolized in man and rabbit by entirely different pathways. In contrast, dogs metabolize the drug in the same general way as man but so slowly that had it been screened in this species, it might well have been discarded.

The evidence strongly suggests that not only do animal tests fail to protect the public, they can actually prove dangerously misleading, adding to the burden of iatrogenic disease. Toxicologists are therefore pursuing an illusion of safety using animals to fulfil political and legal obligations. As if to confirm our suspicions, some drugs are marketed and clinical procedures undertaken despite 'failing' animal tests! The drugs Farlutal and Alexan are both used to treat cancer yet the company literature tells us:

> 'It should be noted that long term administration [of Farlutal] to beagle dogs has resulted in the development of mammary nodules which were occasionally found to be malignant. The relevance of these findings to humans has, however, not been established.'

> (Farmitalia Carlo Erba)

> 'Alexan is known to be teratogenic in some animal species and should be used in women who are or who may become pregnant only after due consideration of the benefit/risk potential.'

> (Pfizer)

Since these drugs are for serious conditions the animal results, it might be argued, could surely be ignored. But if this is true, why perform the experiments in the first place?

Another example is the drug Diane, used to treat acne.[152] Despite causing liver cancer in rats, the drug was still marketed, the company's literature containing the now familiar warning: 'The relevance to humans is unknown.' And when the anti-epileptic drug Tegretol was found to cause liver tumours in rats,[153] the manufacturer's literature stated that the finding constituted '. . . no evidence of significant bearing on the therapeutic use of the drug.' Professor Dennis Parke, a leading British biochemist, provides yet another example, this time the widely used corticosteroid drugs:

> '. . . corticosteroids are known to be teratogenic in rodents, the significance of which to man has never been fully understood, but nevertheless is assumed to be negligible. However, the practice of evaluating corticosteroid drugs in rodents still continues, and drugs which exhibit high levels of teratogenesis in rodents at doses similar to the human therapeutic dose are marketed, apparently as safe, with the

manufacturer required only to state that the drug produces birth defects in experimental animals, the significance of which to man is unknown.'[154]

The same applies to clinical trials and new therapeutic procedures. Scientists at the Burns Research Unit at Birmingham's Accident Hospital carried out trials with an antihistamine drug for its (hopefully) beneficial effects in human volunteers. Preliminary animal tests by other researchers with guinea pigs and rabbits had proved negative but the scientists argued that,

'... negative results with animals do not rule out the possibility that antihistamine drugs may influence the course of human burns. We decided, therefore, to experiment on a group of human volunteers, each of whom would receive burns of the same severity and half of whom would take an antihistamine drug.'[155]

The trial could not be faulted logically because, as other burns researchers point out, '... the biological differences between the various animal species utilized and man are so great that extrapolation of the data to man is fraught with danger'.[156] If the original animal tests were likely to produce such dubious results, why do them in the first place?

But if animal tests are sometimes ignored, they can also be used to imply certain advantages of a company's new product over existing drugs. We have already seen how Opren was promoted as having the potential to modify the disease process in arthritis – an enormous commercial advantage, if true, over other competing drugs that could only alleviate symptoms at best. But artificially induced arthritis in rats is not a good model for the human disease and the results could not be confirmed in patients. On the basis of animal tests another antiarthritis drug Surgam was promoted as giving 'gastric protection', a considerable advantage over similar drugs, which can damage the stomach. But clinical trials showed that Surgam was just like the other drugs and the company, Roussel Laboratories, a subsidiary of Hoechst, were found guilty of misleading advertising and fined £20,000. According to a report in the *Lancet*,[157] expert witnesses for *both* sides '... agreed that animal data could not safely be extrapolated to man.'

In another case the drug company Pfizer was reprimanded by the UK Drug and Therapeutics Bulletin for misleading claims made in advertising the company's antiasthma drug Exirel. On the basis of animal experiments, Pfizer claimed that Exirel had advantages over another similar drug, Salbutamol.[158] The *Drug and Therapeutics Bulletin* called upon Pfizer to,

> 'acknowledge that animal studies proving Exirel "more selective than Salbutamol" are irrelevant to its use in man.'[158]

On the other hand the fact that animal tests are often misleading can form the basis of a company's defence against claims about one of its products. A good example is the Draize eye test and an Ohio court ruling in 1974. A girl had suffered eye damage after using a shampoo. Applying the Draize test the US government's Food and Drug Administration (FDA) found the shampoo was indeed an irritant and went to court against the manufacturers. But the Northern District Court of Ohio ruled in favour of the company partly because the FDA failed to show that results from rabbit eye tests can be transferred to human beings.[159]

In the case of Depo-Provera, a long-acting contraceptive produced by the Upjohn Company, the manufacturers argued that the United States ban on its product should be lifted because animal tests proving it dangerous were misleading.[160] Depo-Provera causes cancer in beagles and monkeys but the company believes neither of these animals provide a good model for assessing the dangers in people. And Britain's Committee on Safety of Medicines, referring to the harmful effects in monkeys, states that the '. . . relevance of this to man has not been established.'[161] Nevertheless, the Minister of Health, disregarding his official committee's recommendations, decided to ban the drug.

So, if animal experiments are misleading, they are at least flexible: they can be deemed inapplicable when necessary, ignored when convenient and used to imply important advantages over competing products. After thalidomide, animal testing increased sharply, partly through the 1968 Medicines Act, yet, paradoxically, it is the very uncertainty of experiments on animals that has led to more and more tests being performed. But, despite increased testing, iatrogenic disease has reached epidemic proportions and drugs continue to be withdrawn because of serious and unforeseen side-effects. Furthermore, major, well-publicized drug disasters such as Eraldin and Opren occurred *after* the Medicines Act was passed, proving that we cannot delegate our responsibilities onto other animals, who only reward us with illusions of safety. Incredibly some argue that more research is necessary to discover *how* to transfer the results from animals to people:

> 'A major goal of modern toxicology is to assess the causes of interspecies differences and susceptibility to toxic agents, with the ultimate hope of trying to improve upon the present capability of predicting human responses based on animal studies.'[21]

It is this obsession with animal experiments that has delayed the development and introduction of humane and more reliable approaches.

In the academic world too, animal experiments have a distinct advantage: when an 'animal model' proves inadequate, another species can be chosen, and another, resulting in almost never-ending scope for scientific publications – the measure of success. The use of animals in experimental psychology is a good example. Scientists at University College Swansea felt that there was little information on how starvation affects fighting between rodents: 'What is available, largely relates to the rat and is equivocal. Because of this a study on the effects of food deprivation in mice ... was undertaken.'[162] And at the University of Newcastle, pigeons have been subjected to brain damage to see the effects on memory.[163] Previous experiments have recorded the effects of similar damage in other species but the scientist felt that '... one species that has not been adequately studied is the pigeon.'

Later on we will see how major drug disasters can be avoided if we adopt other means of assessment and no longer rely on animal experiments. But, even if there were no alternatives, the very fact that vivisection is so dangerously misleading surely argues for new drugs to be strictly limited to those absolutely essential, so that hazards can be minimized. In fact, medical sources reveal that the great majority of new drugs add little or nothing to those available now, being introduced into therapeutic areas already heavily oversubscribed. Millions of animals therefore suffer and die to develop and test products for which there is no real medical need. So let us have a closer look at the industry responsible for this mass carnage – an industry that refers to its products as 'ethical pharmaceuticals.'

1 G. Pickering, *BMJ*, 1615-1619, 26 December, 1964
2 'The Opren Scandal', Panorama, BBC1, 10 January, 1983

3 *BMJ*, 459-460, 14 August, 1982
4 B. Brodie, *Clinical Pharmacology & Therapeutics*, 374-380, volume 3, 1962
5 D. F. Hawkins (Ed.), *Drugs and Pregnancy – Human Teratogenesis and related problems* (Churchill Livingstone 1983)
6 T. Koppanyi and M. A. Avery, *Clinical Pharmacology & Therapeutics*, 250-270, volume 7, 1966
7 N. Irving Sax, *Cancer-causing Chemicals* (Van Nostrand Reinhold, 1981)
8 L. Friedman, *Toxicology & Applied Pharmacology*, 498-506, volume 16, 1969
9 G. Zbinden, *Advances in Pharmacology*, 1-112, volume 2, 1963
10 L. E. Davis, *Journal of American Veterinary Medical Associations*, 1014-1015, volume 175, 1979
11 D. Guthrie, *A History of Medicine* (Nelson, 1945)
12 D. F. Fraser Harris, *New Health*, September 1936: reproduced in reference 149. See also Chapter 5 and references therein
13 L. Whitby, *Practitioner*, 650-660, December, 1937
14 C. T. Dollery in *Risk-Benefit Analysis in Drug Research*, J. F. Cavalla (Ed.) (MTP Press Ltd, 1981)
15 A long – but by no means complete – list of animal models of human diseases is given by C. E. Cornelius, *New England Journal of Medicine*, 934-944, volume 281, 1969
16 *Lancet*, 19, 3 July, 1948
17 See M. J. Ball, et al, *Neurobiology of Ageing*, 127-131, volume 4, 1983
18 *Lancet*, 357-358, 26 February, 1949
19 *Lancet*, 674, 24 March, 1951
20 P. K. Jeffery, et al, *Proceedings of the Physiological Society*, 53P, February, 1984
21 E. J. Calabrese, *Drug Metabolism Reviews*, 505-523. volume 15, 1984
22 M. Massoudi, et al, *Annals of Nutrition & Metabolism*, 26-37, volume 27, 1983
23 *Lancet*, 99, 20 July, 1946
24 See reference 25. Pigs are considered the best available model for human atherosclerosis, but they are regarded as expensive and difficult to work with
25 D. R. Gross, *Animal Models in Cardiovascular Research* (Martinus Nijhoff, 1985)
26 J. P. Whisnant in *Cerebral Vascular Diseases* (2nd Conference), C. H. Millikan (Ed.) (Grune & Stratton, 1958)
27 W. H. Ogilvie, *Lancet*, 419-424, 23 February, 1935
28 W. H. Ogilvie, *Lancet*, 555-560, 21 March, 1953
29 R. Doll, *Cancer*, 2475-2485, volume 45, 1980
30 *Medical Reviews*, 23-35, February, 1951
31 For instance, see *BMJ*, 1732, 12 June, 1982 and *New England Journal of Medicine*, 1226-1232, volume 314, 1986
32 *BMJ*, 1272, 2 December, 1950
33 N. R. Farnsworth and J. M. Pezzuto, paper presented at the University of Panama workshop sponsored by the International Foundation for Science, 1982. Reproduced in *Lord Dowding Fund Bulletin*, 26-34, no. 21, 1984
34 Dr Larry Weisenthal of the Long Beach Veterans Administration

Hospital at the University of California, speaking on BBC Radio 4's 'Medicine Now' programme (4 June, 1985)

35 H. B. Stoner, *Advances in Shock Research*, 1-9, volume 2, 1979
36 L. B. Hinshaw, et al, *Archives of Surgery*, 1410-1418, volume 119, 1984
37 A. Demaria, et al, *Lancet*, 1363, 15 June, 1985
38 See Chapter 5
39 The Wellcome Museum of the History of Medicine (Science Museum, London, 1986)
40 W. Risden, *Lawson Tait, a biographical study* (NAVS, 1967)
41 Lawson Tait, *Transactions of the Birmingham Philosophical Society*, 20 April, 1882, reproduced in reference 40
42 *Birmingham Daily Mail*, 21 January, 1882
43 In reference 149, Dr Beddow Bayly quotes many surgeons who have spoken against the misleading results from animal experiments
44 F. Treves, *BMJ*, 1389, 5 November, 1898
45 *Lancet*, 1003, 5 May, 1951
46 *BMJ*, 659-660, 3 March, 1984
47 R. B. Griepp, et al, in J. S. Najarian and R. L. Simmons (Eds.), *Transplantation* (Lea & Febiger, 1972)
48 *Lancet*, 687-688, 29 March, 1980
49 Dr Albert Iben, Stanford University cardiac surgeon reported in the *Erie Daily Times*, 23 May 1968: reproduced in *Animals Defender*, December, 1968 (NAVS journal)
50 S. W. Jamieson, et al, *Lancet*, 1130-1132, 21 May, 1983
51 C. M. Burve, *Lancet*, 517-519, 8 March, 1986
52 R. Y. Calne, et al, *BMJ*, 1345-1351, 3 December, 1966
53 R. Y. Calne, et al, *BMJ*, 645-651, 14 September, 1963, reports five kidney transplants '. . . in which an attempt has been made to prevent immunological rejection by direct application of experimental methods which had encouraging results with dogs.' Within two months, four of the five patients had died.
54 *Lancet*, 419-420, 22 February, 1986
55 R. Y. Calne, et al, *Lancet*, 1342, 8 June, 1985
56 J. Collier, et al, *Lancet*, 216, 25 January, 1986
57 J. W. Fabre, *Transplantation*, 223-234, volume 34, 1982
58 *Sunday Times*, 5 December, 1982
59 *Guardian*, 25 March, 1983
60 *New Scientist*, 7, 29 November, 1984
61 *Nature*, 88, 8 November, 1984
62 *Daily Telegraph*, 30 June, 1986
63 R. D. Mann, *Modern Drug Use, an Enquiry on Historical Principles* (MTP Press Ltd., 1984)
64 P. F. D'Arcy, et al, *Iatrogenic Diseases* (Oxford University Press, 1979) – recorded in reference 63
65 A. D. Welch in *Drug Responses in Man*, G. Wolstenholme and R. Porter (Eds.) (Churchill Livingstone, 1967). See also reference 129
66 Zomax was withdrawn from the US market in 1983 after five deaths from allergic reactions. In the UK the drug was withdrawn after 23 months and seven deaths and over 500 adverse reactions.
67 M. Weatherall, *Nature*, 387-390, 1 April, 1982

68 G. Zbinden, *Applied Therapeutics*, 128-133, volume 8, 1966
69 G. Zbinden, *The Handbook of Biochemistry and Biophysics* (World Publishing Company of Cleveland, Ohio, 1966)
70 J. T. Litchfield Jnr., *Clinical Pharmacology & Therapeutics*, 665-672, volume 3, 1962
71 A. P. Fletcher, *Journal of the Royal Society of Medicine*, 693-698, volume 71, 1978
72 G. R. Venning, *BMJ*, 199-202, 15 January, 1983
73 F. H. Gross and W. H. Inman (Eds.) *Drug Monitoring* (Academic Press, 1977)
74 *BMJ*, 136-138, 19 July, 1952
75 *Opren: Clinical & Laboratory Experience* (Dista Products Ltd, 26 September, 1980)
76 No mention is made of photosensitivity during animal experiments in reference 75.
77 M. F. Cuthbert in *Current Approaches in Toxicology*, B. Ballantyne (Ed.) (Wright & Sons, 1977)
78 Adverse reactions and human deaths reported in reference 63; original animal tests reported in *Arzneim-Forsch (Drug Research)*, 1100-1107, volume 23, 1973
79 For 1980, 2,524 deaths were officially recorded as caused by use or misuse of drugs (*BMJ*, 223, 15 January, 1983)
80 R. C. Heel, et al, *Drugs*, 169-206, volume 24, 1982
81 *Lancet*, 534, 5 March, 1977
82 R. Hess, et al, *Lancet*, 424-425, 26 August, 1972
83 T. Tsubaki, et al, *Lancet*, 696-697, 3 April, 1971
84 M. H. Briggs in *Biomedical Research Involving Animals*, Proceedings of the XVIIth CIOMS Round Table Conference, Geneva, 8-9 December, 1983, Z. Bankowski and N. Howard-Jones (Eds.) (CIOMS, 1984)
85 The British National Formulary (1983) lists the dose of oral clindamycin as 150-300mg every six hours, that is, a maximum of about 13mg/kg/day. Rats and dogs could tolerate more than 300mg/kg (J. E. Gray, et al, *Toxicology & Applied Pharmacology*, 516-531, volume 21, 1972)
86 *Lancet*, 121, 12 January, 1985
87 J. K. Heiberg and E. Svejgaard, *BMJ*, 825, 26 September, 1981
88 *New Scientist*, 218, 17 July, 1980
89 W. H. Inman in *Monitoring for Drug Safety*, W. H. Inman (Ed.) (MTP Press Ltd, 1980)
90 A. Melville and C. Johnson, *Cured to Death* (New English Library, 1983)
91 F. I. McMahon, *Medical World News*, 168, volume 6, 1965
92 S. J. Alcock, *Proceedings of the European Society for the Study of Drug Toxicity*, 184-190, volume 12, 1971
93 R. Stuart Harris, et al, *Lancet*, 219-220, 28 July, 1984
94 C. Moroni, et al, *Lancet*, 45 and references therein, 7 January, 1984
95 C. F. George, *BMJ*, 1397-1399, 22 November, 1980
96 *Lancet*, 129-130, 19 July, 1980
97 *Anaesthesiology*, 109-110, volume 24, 1963
98 P. D. Stolley, *American Review of Respiratory Diseases*, 883-890, volume 105, 1972
99 Reference 101 states that doses of 1mg/kg did not kill cats. This is 175

times the dose subsequently found to be dangerous to asthmatics, that is, 0.4mg for a 70kg person or 0.0057mg/kg

100 S. Carson, et al, *Pharmacologist*, 272, volume 18, 1971
101 J. M. Collins, et al, *British Journal of Pharmacology*, 35-45, volume 36, 1969
102 I. Rosenblum, et al, *Toxicology & Applied Pharmacology*, 1-8, volume 7, 1965
103 See references 9 and 104
104 R. Levine, *Pharmacology: Drug Actions & Reactions* (Little, Brown and Co., 1978)
105 Estimate by Dr Sidney Wolfe, director of the Ralph Nader Health Research Group (*Lancet*, 353, 11 February, 1984)
106 D. H. Chatfield and J. N. Green, *Xenobiotica*, 133-144, volume 8, 1978. Reference 2 shows that metabolism of Opren is far slower in elderly patients than in younger ones.
107 D. V. Parke and R. L. Smith (Eds.), *Drug Metabolism – from Microbe to Man* (Taylor & Francis, 1977)
108 Report on the LD50 Test, Advisory Committee on the Administration of the Cruelty to Animals Act 1876 (Home Office, 1979)
109 G. E. Paget (Ed.), *Methods in Toxicology* (Blackwell Scientific Publications Ltd, 1970)
110 *Human Toxicology*, 85-92, volume 3, 1984
111 G. Zbinden and M. Flury-Roversi, *Archives of Toxicology*, 77-99, volume 47, 1981
112 Sixth Report of the Cosmetic Ingredient Review Expert Panel, *Journal of the American College of Toxicology*, volume 3, 1984
113 *Clinical Toxicology of Commercial Products*, 1976
114 D. Salsburg, *Fundamental & Applied Toxicology*, 63-67, volume 3, 1983
115 R. Heywood in *The LD50 Test: Evidence for Submission to the Home Office Advisory Committee*, prepared by CRAE, the Committee for the Reform of Animal Experimentation, August, 1977
116 V. K. H. Brown, *ibid*
117 M. Dawson, *Cellular Pharmacology* (Charles C. Thomas, 1972)
118 Reported in reference 111
119 R. Goulding in *Monitoring for Drug Safety*, W. H. Inman (Ed.) (MTP Press Ltd, 1980)
120 R. H. Jackson, et al, *BMJ*, 1468, 12 November, 1983
121 P. Rogers in *The LD50 Test*, see reference 115
122 D. Lorke, *Archives of Toxicology*, 275-287, volume 54, 1983
123 *New Scientist*, 23 June, 1977
124 A. D. Dayan, et al, *Lancet*, 555-556, 10 March, 1984
125 R. Bass, et al, *Archives of Toxicology*, 183-186, volume 51, 1982
126 Home Office Statistics, 1986 (HMSO, 1987)
127 R. Heywood in *Testing for Toxicity*, J. W. Gorrod (Ed.) (Taylor & Francis, 1981)
128 B. Ballantyne and D. W. Swanston in *Current Approaches in Toxicology*, B. Ballantyne (Ed.) (Wright & Sons, 1977)
129 F. Coulston and D. Serrone, *Annals of the New York Academy of Science*, 681-706, volume 162, 1969. See also references 128 and 130
130 E. V. Buehler and E. A. Newmann, *Toxicology & Applied Pharmacology*,

701-710, volume 6, 1964

131 C. Muir, et al, *Toxicology Letters*, 1-5, volume 18, 1983

132 R. B. Kemp, et al, *Cytobios*, 153-159, volume 36, 1983

133 D. W. Swanston in *Animals and Alternatives in Toxicity Testing*, M. Balls, R. J. Riddell and A. N. Worden (Eds.) (Academic Press, 1983)

134 C. S. Weil and R. A. Scala, *Toxicology & Applied Pharmacology*, 276-360, volume 17, 1971

135 *In vitro* alternatives to the Draize test described at an International Workshop in Switzerland (April, 1984). Proceedings published in *Food & Chemical Toxicology*, no. 2, volume 23, 1985

136 F. J. Di Carlo, *Drug Metabolism Reviews*, 409-413, volume 15, 1984

137 A. D. Dayan and R. W. Brimblecombe (Eds.), *Carcinogenicity Testing* (MTP Press Ltd, 1978)

138 R. Howe, *Nature*, 594, 7 August, 1965

139 *The Sunday Times* 'Insight' Team, *Suffer The Children – The Story of Thalidomide* (Andre Deutsche, 1979)

140 T. H. Shepard, *Catalogue of Teratogenic Agents* (John Hopkins Press, 1976)

141 Reproduced in references 5 and 63

142 K. S. Larsson, et al, *Lancet*, 439, 21 August, 1982

143 *Handbook of Teratology*, volume 4, J. G. Wilson and F. C. Fraser (Eds.) (Plenum Publishing Co. Ltd., 1978)

144 P. Lewis in reference 5

145 R. W. Smithells in *Monitoring for Drug Safety*, W. H. Inman (Ed.) (MTP Press Ltd, 1980)

146 A. McGlashan, *Lancet*, 1332-1333, 24 December, 1955

147 'The Discovery of Penicillin', BBC Radio 4, 5 August, 1981

148 H. Florey, *Conquest,* January, 1953 – reproduced in reference 149

149 M. Beddow Bayly, *The Futility of Experiments on Living Animals* (NAVS, 1962)

150 Reference 149 and references therein

151 F. E. Karch in *Side Effects of Drugs*, Annual 3, M. N. G. Dukes (Ed.) (Excerpta Medica, 1979)

152 Diane promotional literature, Berlimed Pharmaceuticals, a Division of Schering Chemicals Ltd

153 Tegretol promotional literature, Geigy Pharmaceuticals Ltd

154 D. V. Parke in *Animals in Scientific Research: An Effective Substitute for Man?*, P. Turner (Ed.) (Macmillan, 1983)

155 S. Sevitt, et al, *BMJ*, 57, 12 July, 1952

156 C. L. Fox Jnr. and S. E. Lasker, *Annals of the New York Academy of Sciences*, 611-617, volume 150, 1968

157 J. Collier and A. Herxheimer, *Lancet*, 113-114, 10 January, 1987

158 *SCRIP*, 18, 10 September, 1984

159 D. Pratt, *Alternatives to Pain* (Argus Archives, 1980)

160 *Washington Post*, 11 January, 1983

161 *BMJ*, 1426, 15 May, 1982

162 S. Al-Malike and P. F. Brain, *Animal Behaviour*, 562-566, volume 27, 1979

163 A. Sahgal, *Behavioural Brain Research*, 49-58, volume 11, 1984

164 By 1980 ICI had received 2,600 claims; payments had been made in 1,200 cases with 1,000 being rejected. ('A Question of Balance', Office of

Health Economics, 1980)
165 R. Cockel, *Gut*, 515-518, volume 28, 1987
166 'Agenda for Health', A Report from the Association of the British Pharmaceutical Industry, 1987
167 'World in Action', Granada Television, 9 November, 1987
168 *SCRIP*, 2, 2 October, 1987
169 D. Schuppan, et al (Eds.), *The Contribution of Acute Toxicity Tests to the Evaluation of Pharmaceuticals* (Springer Verlag, 1986)
170 G. N. Volans in reference 169
171 R. E. Davies, et al, *Journal of the Society of Cosmetic Chemists*, 371-381, volume 23, 1972

CHAPTER 4

Keep taking the tablets

'Most of the tens of thousands of drugs on the world market are either unsafe, ineffective, unnecessary or a waste of money.'

(Health Action International, 1986[1])

Drugs are big business. In 1985 the world's 12 best-selling drugs achieved sales of over $6 billion.[2] Companies exist primarily to make a profit, to satisfy their shareholders. And they are extremely successful. In 1983 the 25 leading companies reported sales of $35 billion.[3]

Company	Pharmaceutical Sales ($million)	Country of origin
1. Hoechst	2,628	West Germany
2. Bayer	2,452	West Germany
3. Merck & Co	2,217	USA
4. American Home Products	2,144	USA
5. Ciba-Geigy	2,053	Switzerland
6. Pfizer	1,694	USA
7. Eli Lilly	1,535	USA
8. Hoffman La Roche	1,511	Switzerland
9. Sandoz	1,419	Switzerland
10. Bristol-Myers	1,360	USA
11. Smith Kline	1,339	USA
12. Abbott	1,299	USA
13. Takeda	1,291	Japan
14. Warner-Lambert	1,286	USA
15. B. Ingelheim	1,213	West Germany
16. Upjohn	1,213	USA
17. Johnson & Johnson	1,118	USA
18. Glaxo	989	UK

Company	Pharmaceutical Sales ($million)	Country of origin
19. E. R. Squibb	978	USA
20. Rhone-Poulene	895	France
21. Am Cyanamid	883	USA
22. Schering-Plough	882	USA
23. I.C.I.	839	UK
24. Wellcome	836	UK
25. Beecham	781	UK

Profit margins are also huge. In 1984/5 Eli Lilly had sales over $2 billion with profits of $669.1 million, a profit margin of 32.98 per cent. ICI were not far behind with a profit margin of 30.89 per cent based on pre-tax profits of $332 million and sales of over $1 billion.[4]. With so much at stake, the choice of products to be developed is likely to depend far more on marketing considerations than actual medical need. Although potential sales seem large, relatively little is spent on research into Third World diseases because few developing countries can afford to pay for expensive medicines. In its neglect of the health problems of poorer countries, the pharmaceutical industry is not alone: only 1 per cent of all money spent by industry, governments and charities on medical research is devoted to the major diseases of developing countries.[5]

Research and development then is mainly targeted at diseases or conditions that are common in affluent societies such as arthritis, asthma, obesity, anxiety and depression, high blood pressure and other cardiovascular problems. In the UK over 20 million people experience some form of rheumatic complaint and 8 million visit their doctors – 23 per cent of all consultations![6] An incredible 15-20 per cent of all prescriptions are for mood-changing drugs such as antidepressants, sleeping pills, sedatives, stimulants and tranquillizers.[7] Valium alone had world sales of £250 million in 1984.[8] Dr Richard Rondell, Director of clinical research at Bristol-Myers, states:

> 'If I look at the marketing activities of many of the major companies the trend is to go for large markets such as the treatment of hypertension [high blood pressure] and to say that even if we get one or two per cent of that market that is probably the best way to get a return on investment.'[9]

And since many drugs only treat symptoms and neglect the underlying causes, patients carry on taking the tablets, thus boosting profits still further.

The promotion of unsuitable, even dangerous products especially in Third World countries, shows once again that profits come well before people's health. In 1982 the *International Journal of Health Services* reported an investigation into the promotional activities of companies in the UK, the USA, Latin America, Africa and Asia.[10] In contrast to the promotional material used in the United States and Britain, literature provided for doctors in Third World countries was marked by gross exaggeration of product effectiveness and minimized or completely omitted potential hazards. Much of the promotion concerned 'luxury products' such as costly tonics and appetite stimulants, marketed in poor countries where the pressing need is for food. The survey also found that bribery of influential physicians and key government officials may play an important role in irrational drug promotion and use in the Third World.

One example quoted by the report is chloramphenicol, an antibiotic restricted to very serious conditions in advanced countries because it can cause fatal blood disease. Yet in Indonesia only six out of nearly 30 brands containing chloramphenicol gave any warning of possible blood damage. Another case is clioquinol, used to treat diarrhoea and marketed by Ciba Geigy under the name of Enterovioform. Although removed from the market in several industrialized countries, it was still being promoted by many companies in the Third World despite causing at least 10,000 Japanese cases of SMON (subacute myelo-optic neuropathy), a disease characterized by irreversible paralysis, visual disturbances and even blindness. Yet six products containing clioquinol in Malaysia/Singapore and three in the Phillipines, and marketed by a variety of companies, were being promoted without warnings. Oxfam describe how the painkilling drug amidopyrine was withdrawn from the British market in 1963 because it can lead to a fatal blood disease called agranulocytosis.[11] In 1977 Ciba Geigy and Sandoz, who marketed several amidopyrine-based products, anounced their intention to remove the drug by the end of the year. Three years later, Oxfam report, Ciba products containing amidopyrine were still being sold in ten developing countries and in only one case was any warning given. In November 1981

evidence appeared in the *Lancet* that since 1977 Ciba Geigy had
continued to manufacture drugs containing amidopyrine and
had been selling off their old stocks.[12]

An Oxfam analysis of products marketed by the subsidiaries
of two leading British firms with factories in Bangladesh reveals
a range top-heavy with drugs not relevant to priority needs.[11] Of
Fisons 31 products listed as available in Bangladesh during
1981, only four were included in the World Health Organisation's list of around 200 essential drugs whilst just 14 of Glaxo's
products were listed by the WHO. In fact 22 (or nearly 40 per
cent) of the Glaxo range were vitamins and tonics whilst only
three of these brands were available in the UK.

The actions of many companies in the Third World can only
add to health problems that are already appalling. But even in
the West, company propaganda can be false or misleading[13]
whilst drug side-effects are reproduced in such small print that
in one study doctors were forced to use a magnifying glass![14]

Pharmaceutical companies make money in two ways: by
promoting older, well-established drugs under their own brand
name and by developing new chemical substances that can then
be patented. To sell a drug like aspirin at a good price it would
have to be given a brand identity by calling it something else, for
example ASPRO. Unfortunately the National Health Service
spends millions of pounds every year on expensive brands of
medicines instead of buying unbranded versions of the identical
drug, which are just as good. For instance, the active ingredient
of Valium is diazepam, which is referred to as the generic
version. If the NHS were to buy 1,000 5 mg tablets of diazepam,
they would cost £1.50; Valium, the identical drug, but with a
brand name, costs £13.62![15] A switch to generic prescribing, as
recommended by the Greenfield Committee, would save the
NHS tens of millions of pounds.[16]

Old drugs can also be revamped by changing their
formulation or by combining them with others. This adds to the
company's product range and gives them something new with
which to attract doctors' attention. And if products differ from
each other even slightly, the likelihood of price competition is
reduced, allowing the companies to set their own prices instead.
Over a third of all newly marketed products are combinations,
yet they are not usually considered ideal on medical grounds –
only seven of the WHO's list of essential drugs are
combinations.[16]

The alternative is for companies to develop entirely new chemical substances called 'new chemical entities' (NCEs). This is what all major firms try to do because then they have patent protection, which gives them monopoly rights on their inventions for up to 20 years, although by the time the drug is marketed this can be reduced by half. As we shall see, the great majority of NCEs are minor molecular modifications of older and well-established products, and therapeutically there is little to choose between them.[16] Such 'me-too' drugs have no great medical advantages but they are a safe investment: by making only slight chemical changes to the original drug, the new product is likely to have similar therapeutic effects but with the considerable advantage that it can now be patented. Innovation, then, usually involves making me-too drugs that bring commercial rather than medical gain. According to Dr J. W. Dunne of the World Health Organisation, the costs of drug development mean that,

> '... more and more drugs will be introduced that are substitutes for those already on the market in order that a patented product should remain available.'[17]

As newly patented products, me-toos are likely to cost a lot more and this, together with a preference for expensive brand names where cheaper generic versions are already available, means that the NHS receives poor value for money. Whilst millions are squandered to buy unnecessarily expensive drugs, less than 0.4 per cent of the nation's health budget goes on prevention where the cash is really needed.[19] Vast sums are also spent on ineffective medicines. In 1967 an official enquiry into the 'Relationship of the Pharmaceutical Industry with the National Health Service', chaired by Lord Sainsbury, announced that a third of the most prescribed drugs were 'undesirable preparations.'[16] World-wide, a detailed report by Health Action International documents an appalling waste of resources:

- of antidiarrhoeals , 80 per cent have no proven effectiveness in the treatment of acute diarrhoea
- out of 546 products on the market for coughs and colds in five areas of the world, 456 (83.5 per cent) are irrational combinations
- more than three-quarters of the 888 vitamin preparations on the market in the same areas are either irrational, ineffective, or the incorrect dose and could not be recommended for use

- three-quarters of the 356 analgesics on the market should not be recommended for use because they are dangerous, ineffective, irrational or needlessly expensive
- of the 217 non-steroidal anti-inflammatory drugs (NSAIDs) on the world market, 73 per cent could be removed because of their poor safety records, their lack of significant advantage and their much higher cost over safer products.[1]

Whatever the product, the key to success is promotion and the pharmaceutical industry is notorious for its aggressive high-pressure sales techniques. In 1982 the drug industry in the UK spent £150 million on promotion, roughly equivalent to £4,000-£5,000 for every general practitioner.[16] In the UK promotional expenditure, largely funded by the NHS, is about 11-15 per cent of sales but in the United States, Italy and West Germany, it runs as high as 22 per cent.[20]

The companies promote their products by advertising in medical journals, direct mailings, sponsored meetings and through an extensive chain of detail men or medical representatives. Considerable attention is given to consultants, the 'opinion formers', because general practitioners are so influenced by their recommendations.

In 1982 an Italian drug company, Carlo Erba, invited 100 senior British rheumatologists to a four-day symposium in Venice to mark the UK launch of its new anti-arthritis pill Flosint. Participants travelled on the luxurious Orient Express and stayed in a luxury hotel. On board the train BBC TV journalist Tom Mangold interviewed one of the rheumatologists in the middle of a five-course meal:[22]

'Is this trip costing you anything at all?'

'It is costing me money, yes.'

'And . . . what do you have to spend money on? . . . The travel is free?'

'The travel isn't totally free, no. I paid my own fare to the airport.'

'The accommodation is free?'

'The accommodation is free.'

'The food is free?'

'The food is free . . . the drink is not free.'

Attempts had been made to stop the TV crew travelling on the train but they had taken the precaution of booking extra seats under false names. A year later, in 1983, Flosint was withdrawn from the British market on safety grounds:[23] its bid for a slice of the lucrative market in arthritis pills had failed.

The UK launch of Opren was even grander, being held aboard a luxury cruise ship on the Rhine, and a symposium on the drug held in Paris a year later was estimated to have cost Eli Lilly and its UK subsidiary Dista, around £250,000.[22] During the launch it was revealed that just one drug representative had spent over £800 on food and drink in 13 days, entertaining GPs and their wives!

Other effects are more subtle. Industry tries hard to form a close relationship with doctors and scientists, providing financial support for scientific meetings and research. Not only is this good public relations but the research will be of direct benefit to the company. But, as doctors become more dependant on industry, they may be less inclined to condemn its more corrupt activities. Medical education has come to rely on industry for funding and, according to clinical scientist Michael Rawlins of Newcastle University:

> 'Postgraduate education is thus tending to become the responsibility of the drug industry rather than of post-graduate deans, clinical tutors and the [medical] profession itself. This trend should be a cause of major concern to us all, because of the potential for distorting postgraduate medical education away from the needs of patients and the health service, towards the requirements of industry.'[24]

Professional journals are also heavily dependant on drug company advertising. In the late 1970s the *Journal of the American Medical Association* received $7 million a year from drug advertising,[21] but how does this affect editorial policy?

We have already seen how drugs made relatively little contribution to the vast improvements in health over the past hundred years and how they seem unable to cope with today's epidemics of heart disease and cancer. But, whilst some medicines have proved valuable, what of the rest, basking in the glory of the few? In America, where the government's Food and Drug Administration regulates the introduction of new drugs, products are rated according to their therapeutic value. An analysis of 1,935 drugs up to April 1977 revealed that only 3.3 per cent constituted important therapeutic gains whilst just 15.5 per cent represented 'modest' advances.[20] The vast majority (79.4 per cent) afforded 'little or no gain.'

Unfortunately the UK's drug watchdog, the Committee on Safety of Medicines, does not assess new drugs in the same way but, nevertheless, medical surveys show a similar pattern to that found in the United States. About 80 per cent of newly marketed products are new formulations, combinations or duplications of *existing* drugs[18] and, as we have seen, are often introduced for commercial reasons, to give old products a new lease of life. Yet they are still tested on animals.

A recent tragic example of a new formulation is Osmosin, a slow-release form of the well-established drug indomethacin, used to treat arthritis. Unfortunately Osmosin proved even more hazardous than indomethacin, causing 40 deaths in the UK alone, and was withdrawn late in 1983 after just ten months.[25]

The remaining 20 per cent of drugs introduced each year are the new chemical entities (NCEs), so it is here we should look for any real medical advances. But a survey by the Medicines Division of the Department of Health and Social Security (DHSS) revealed that even these,

'. . . have largely been introduced into therapeutic areas already heavily oversubscribed' and

'. . . for conditions which are common, largely chronic and occur principally in the affluent Western society. Innovation is therefore largely directed towards commercial returns rather than therapeutic needs.'[26]

The survey found that between 1971 and 1981, 204 NCEs were introduced onto the UK market, mostly in areas already well supplied. These were mainly anti-inflammatory agents (for arthritis), minor tranquillizers, antidepressants, antibiotics, beta-blockers and other drugs to treat high blood pressure. Only one drug had been introduced for the world's commonest disease, schistosomiasis, which is prevalent in many Third World countries. Apart from drugs like aspirin and phenacetin, which also have anti-inflammatory properties, seven other non-steroidal anti-inflammatory agents were available in 1971. Ten years later a further 21 had been introduced. In 1971, three beta-blockers (used to treat heart problems and high blood pressure) were marketed in the UK; by 1981 a further ten had been licensed.

Another survey, conducted by the prestigious *Drug and Therapeutics Bulletin* together with *Medical Letter*, went a step further and ranked the products according to their therapeutic value.[27] Of 107 NCEs marketed in the UK between 1971 and 1976, only about 8 per cent were classed as important gains whilst around 70 per cent were thought to offer little or no advantage over existing products. A similar picture emerges for the world drug market. An analysis by health economist Etienne Barral found that over 70 per cent of NCEs introduced between 1975 and 1984 offered no therapeutic improvement over existing products.[28] And once again most products were introduced into therapeutic areas already well supplied. There were 19 antidepressants, 21 drugs to treat anxiety, 19 beta-blockers, 52 antibiotics and 56 anti-inflammatory agents. One third of the products offering no improvement still proved a commercial success – a testimony to the companies' aggressive advertising and promotional activities. Yet the industry's international code of practice still claims that all products should, '. . . have full regard to the needs of public health.'[16]

According to British statistics,[29] most animals are used to develop new drugs and appliances (1,586,075 or 51 per cent of the total in 1986), yet the vast majority clearly suffer and die to test products for which there is no real medical need. Dr R. D. Mann, Principal Medical Officer at the DHSS, argues that discouraging the development of me-too or copycat drugs '. . . would save many animal lives.' Such products, he says,[25] should not be reimbursed by the NHS, which would ultimately deter companies from producing them. A number of health and consumer groups such as Health Action International[1] and Social Audit[16] support the idea of a rational drugs policy based on the WHO's concept of essential drugs. By discouraging or preventing the development of me-too drugs, an essential drugs policy would also benefit doctors, patients and the Health Service, which would have more cash for other areas such as disease prevention. With so many closely related drugs it is impossible for doctors to fully appreciate the dangers, so fewer medicines would facilitate more in-depth knowledge. There would be fewer bad drugs and less confusion about which medicine to choose – both strategies that would lead to a reduction in iatrogenic disease. Britain's highly respected *Drug and Therapeutics Bulletin* agrees that the use of a smaller number of medicines should increase knowledge of their real benefits and hazards, so leading to safer prescribing.[49] The *Bulletin* also states that:

'The existence of many apparently similar preparations seldom increases therapeutic options, but greatly increases the risk of unwanted effects and makes them more difficult to detect.'[49]

Companies contribute to bad prescribing habits, and hence the high level of iatrogenic disease, partly by making undesirable products, partly by making too many of them and partly by the way they promote them.[16] Doctors cannot possibly choose between thousands of heavily promoted drugs, each with reams of propaganda extolling its virtues, nor can they properly assess the dangers. Almost inevitably they will be slow to react, if only because of the great mountain of literature they receive. For instance, ICI's heart drug Eraldin, which caused deaths and severe eye problems including blindness, was widely prescribed right up to the time it was withdrawn in the mid-seventies, despite frequent warnings.[16] Apart from increasing the level of iatrogenic disease, over-prescribing can have other serious effects.

When antibiotics were first developed they were seen as magic bullets that would radically change the treatment of infectious disease. Now, because they are prescribed for trivial illness and for diseases that do not respond to antibiotics, resistant, more virulent strains of bacteria have arisen that pose a real threat to health. The WHO says that '. . . the problem is global and is the result of widespread and indiscriminate use of antimicrobial drugs in man and animals.'[30]

About half of all antibiotics produced are administered to animals reared for food to prevent disease and promote growth.[1] Factory-farmed animals are kept in stressed and overcrowded conditions, so facilitating the spread of disease should one animal become infected. By dosing them with antibiotics, the argument goes, outbreaks of disease can be prevented. Unfortunately resistant bacteria can then pass to humans.

The massive market for antibiotics, estimated at $15 billion annually, is obviously a major factor behind the misuse – since the drug industry is profit-orientated, it tries to increase the sales of antibiotics. It has been pointed out that '. . . powerful marketing techniques used by the pharmaceutical industry often undermine appropriate prescribing habits',[1] so it could be argued that the industry itself is partly responsible for the problem that, predictably, it is trying to solve by developing new, more effective antibiotics to beat the 'super bugs'. Since misuse is the real problem, it can only be solved by rational prescribing (and the elimination of factory farming): more new drugs will not change the way they are prescribed.

In Norway, new medicines are only introduced if there is judged to be a *medical need* for them. Compared to the UK's 18,000 licensed drugs, containing 3,000 active ingredients,[31] Norway has just 1,900 formulations based on a list of 800 specific chemical entities,[32] itself three to four times the World Health Organisation's list of essential drugs.[33] Norway's smaller number of drugs seems to have no adverse effect on life expectancy. Indeed Norwegians can expect to live about one-and-a-half to two-and-a-half years longer than their counterparts in the UK. The British Medicines Act, however, specifically states that the *comparative* efficacy of a new medicine shall not be taken into account when considering product license applications. In other words unnecessary, me-too drugs can be marketed even though they offer no real advantage over existing products. Flosint, introduced onto the British market in 1982, became the twenty-third non-steroidal anti-inflammatory drug (NSAID), all of them competing for a slice of the £100 million market in anti-arthritis pills.[22]

By 1987 the UK market for NSAIDs had reached £150 million and was growing at a rate of 12 per cent annually.[2] The world market is worth £2.5 billion. One doctor, writing satirically in the *Lancet*, described NSAIDs as, ' "new sorts of aspirin in disguise" or NSAID for short.'[34] And in 1983 when another NSAID, Zomax, was withdrawn after deaths from severe allergic reactions, the *Guardian* commented editorially:

'Zomax and Opren were both non-steroidal anti-inflammatory drugs. There are more than 20 such drugs on the market, all of which are used in arthritis and related conditions. Zomax was marketed in 1980. Why was it thought necessary to market such a new drug when there were at least 20 others of a similar nature? One point of view . . . is that, although Zomax was one of this general family of drugs, it was chemically different from the others, and it was legitimate to market and prescribe slightly different substances on the grounds that they might help patients where other drugs from the same family had failed.

'But another view, put to us by a leading pharmacologist, revived the original doubts. Zomax may be chemically different from the rest of the anti-inflammatory family, but pharmacologically there is no difference at all. In other words, when it comes to treating patients, there is little practical difference between them all. Zomax, however, was licensed to treat postoperative pain, for which the others had no license; after Zomax came on the market, various others duly appeared also to treat postoperative pain. So the point about Zomax is not that it was different from the others but that it was cleverly used to plug a gap that existed in the market. And why did doctors choose it? Because of the way it was promoted and advertised. Why can't doctors, instead of relying on drug industry promotions, depend upon a list of drugs in each therapeutic class, in which the top one is most recommended by virtue of its proven worth and known side-effects, the second one on the list is the next most trustworthy, and so on? Might this not be one way of switching off the commercial drug escalator which so palpably fails the test of patient safety?'[35]

Eventually, and if only for economic reasons, the British government decided to limit the number of drugs in certain less essential therapeutic areas, to a few cheaper versions that could be prescribed on the NHS. The scheme covered seven categories: laxatives, tonics, cough and cold remedies, antacids, vitamin preparations, sedatives and tranquillizers, and analgesics for mild to moderate pain. The move came in a desperate attempt to cut the enormous NHS drug bill that in 1984 stood at £1.4 billion. By 1985, and after considerable pressure, the list had been extended from 30 to 108 items that could be reimbursed by the NHS. Even so, as many as 1,800 drugs in just seven therapeutic categories were now deemed unnecessary and no longer available on the NHS.[36] As the *Lancet* pointed out,

'The advisor's selection of allowable drugs has vividly demonstrated that a vast number of preparations in these categories are in fact unnecessary.'[37]

On medical grounds the *Lancet* considered the scheme sound[37] although some doctors objected because, they felt, it interfered with their clinical freedom. But clinical freedom can become an excuse for bad medicine as some doctors prove unable to prescribe efficiently. On the other hand, an analysis of the *financial* impact led stockbrokers W. Greenwell & Co. to conclude that 'The restricted drugs list . . . could well represent the thin edge of a very dangerous wedge',[38] which explains the industry's almost hysterical opposition to the plan.

The Association of the British Pharmaceutical Industry, the drug industry's trade association, staged a massive public

relations campaign against the proposals, no doubt worried that
they might be extended to other therapeutic areas. In fact, some
hospitals already operate a form of voluntary restricted list.
Ninewells Hospital in Dundee uses six of the more than 20
available NSAIDs[38] whilst the World Health Organisation lists
just three, including aspirin.

If patient safety and medical need are to come before the
industry's huge profit margin, it is now time for the government
to bring the total list of drugs prescribable on the NHS into line
with the WHO's concept of essential medicines. New products
could still be added to the approved list, but only if they have
significant clinical advantages and represent a real medical need.
According to the consumer group Social Audit, in a critical
guide to ineffective, inappropriate or extravagantly prescribed
drugs:

> 'We do not need and cannot afford most of the medicines we
> now use, and should therefore be urgently thinking in terms
> of a national medicines policy based on the WHO's concept
> of essential drugs. These are drugs which are ". . . restricted
> to those proven to be therapeutically effective, to have
> acceptable safety and to satisfy the health needs of the
> population." '[16]

But if the industry is concerned about essential drug lists it must
be just as worried about the growing interest in alternative
medicine. Even the highly orthodox *British Medical Journal*
recently acknowledged that 'One of the few growth areas in
contemporary Britain is alternative medicine',[39] with the
number of practitioners growing by about 11 per cent a year.[40]
In one study 33 per cent of patients with rheumatoid arthritis
and 39 per cent of those with backache admitted to their doctors
that they had consulted an alternative practitioner.[39] And a
recent Consumers' Association survey found that alternative
medicine improves the health of four out of five people who have
tried it.[41] More than 80 per cent had turned to alternative
therapies – including osteopathy, homoeopathy, acupuncture,
chiropractic and herbalism – after being dissatisfied with
conventional treatment. Why should this be?

Apart from a high incidence of side-effects, many drugs are
unsatisfactory in the long term because they only treat symptoms
and neglect the underlying causes. This is often an inevitable
consequence of the way in which they are developed. For
instance, as there is no adequate animal model of arthritis,[42]

experimenters deliberately induce the symptoms (pain and inflammation) in animals and test new drugs for their effect on these. This in turn leads to drugs that, at best, reduce the pain and inflammation so often associated with the human disease but which neglect the underlying causes. Furthermore, for patients with osteoarthritis, anti-inflammatory drugs might actually make things worse simply by killing pain.[43]

The function of pain is to keep people from putting undue stress on their arthritic joints so that damage can be healed by the body's natural defences. By reducing pain, patients can walk around more on their arthritic joints, contributing to the damage. By eliminating the protective effect of pain the drugs may actually be making things worse.

The development of the anti-ulcer drug Tagamet (and later its rival Zantac) was hailed as an important breakthrough by orthodox medical sources and the drug does provide an alternative to surgery, at least in the short term. Unfortunately, although the ulcer heals, the patient is not cured and when treatment is stopped, most become ill again: in 85 per cent of cases the ulcer returns.[44] As a result the drug may be given for prolonged periods, perhaps even for life, which is certainly good for profits but not necessarily for patients as the risks of side-effects increase. Once again symptoms have been controlled, for a time, but the underlying cause has been ignored. And the same is true for many other conditions such as asthma, anxiety and depression, allergies and high blood pressure. By neglecting the underlying causes the net result is that illness is perpetuated. The idea that disease is merely a set of symptoms may well suit industry (Tagamet was the world's best-selling drug[8] with sales of £670 million in 1984), but patients are becoming increasingly disillusioned, turning more and more to alternative forms of treatment.

Apart from a notable lack of side-effects compared with orthodox drugs, alternative or complementary medicine has two main advantages. First of all, therapists treat the whole person rather than isolated symptoms and cure their patients by finding, and then eliminating, the root cause of disease. A classic example is backache where GPs often prescribe pain-killing drugs for long periods. The osteopath or chiropractor will, instead, seek out the cause and often cure the patient. Furthermore alternative practitioners bring to medicine a personal touch that means so much to people. We are all familiar with the hasty three-minute consultation, the doctor's pen poised above the prescription pad as our symptoms are described. Consultations with alternative therapists take far longer, with diet and lifestyle factors often being discussed, allowing patients to feel valued and actually part of the healing process.

The growing popularity of alternative medicine has, at last, forced doctors and the medical establishment to take the issue more seriously. Responses have ranged from the predictably hostile to a more open attitude, recognizing that alternative treatments can make a valuable contribution. A recent survey of general practitioner trainees gives some insight into the status of alternative medicine.[40] Out of 86 doctors responding to the survey, 18 were using at least one alternative method themselves with 70 wanting to train in one or more. Of these 86 doctors, 31 of them had referred patients for such treatments. The most commonly used alternatives were hypnosis, manipulation, homoeopathy and acupuncture and, in view of the present emphasis on orthodox medicine, the results were encouraging. Nevertheless, there were some astonishing blind spots, as the tables below reveal. For instance, 50 per cent of the doctors had never heard of naturopathy even though this is the most basic of all forms of medicine, relying mainly on diet.

Naturopathy, or nature cure, is based on the idea that the body can cure most illness itself, given the proper conditions in which to do so. These include eating only natural and wholesome foods, regular exercise, exposure to sunlight, frequent bathing and occasional fasting to clear poisons from

Knowledge of some alternative therapies among 86 general practitioner trainees (Figures are numbers of doctors)

	Never heard of	Heard of only	Know something of	Know a lot
Hypnosis	—	12	64	10
Acupuncture	—	18	64	4
Homoeopathy	—	35	49	2
Osteopathy	1	45	39	1
Herbalism	8	62	16	—
Naturopathy	43	39	4	—

the system, all of which stimulate the body's powers of self-healing. The greatest responsibility in naturopathy lies with the patient who is often required to change long-standing lifestyle and dietary habits.

Attitudes of 86 general practitioner trainees to some alternative therapies (Figures are numbers of doctors)

	Useful	Useless	Don't know
Acupuncture	76	2	8
Hypnosis	74	1	11
Homoeopathy	45	3	38
Osteopathy	39	5	42
Faith Healing	27	15	44
Herbalism	13	7	66
Naturopathy	—	5	81

Some alternative therapies have a long history: the ancient Chinese system of acupuncture has been practised for over 2,000 years but only recently accepted in the West. Even today Chinese patients undergoing major surgery are given the chance of acupuncture anaesthesia. Other systems, such as homoeopathy, osteopathy, chiropractic and the use of hypnotism, all became firmly established during the nineteenth century.

The oldest and most widely practised form of medicine in the world is herbalism – the use of herbs and plants to treat disease. Like other alternative therapies, herbal medicine is not based on animal experiments, being tailored to the patient's individual requirements. Research by herbal doctors and manufacturers relies on long-established use and experience in practice: knowledge accumulated during past generations means that history has been one long clinical trial!

Until recently nearly all medicines were based on plants and many of today's drugs still are: the powerful painkiller morphine is derived from poppies, the heart stimulant digitalis comes from the foxglove, the antimalarial drug quinine comes from the bark of the cinchona tree and the herb qinghao provides another treatment for malaria.[45] Qinghao has been used in China as a

medicine for 2,000 years and was recommended for malaria in
AD341. Now trials have shown that qinghaosu, an extract of
qinghao, is indeed effective in treating malaria, being less toxic
than chloroquine. Curare is derived from wourali root, which
was first used by South American indians as a paralysing arrow
poison. One of its active alkaloids is now used as an important
muscle relaxant in modern surgery.[11] The rose periwinkle plant
(vinca rosea) is the source of the modern drug vincristine, used
to treat leukaemia. In fact, it has been estimated that more than
half the prescriptions written by American physicians contain
plant-derived medicine – that is, a drug that has either been
extracted from a plant or one that has been synthesized to
duplicate or improve on a plant substance.[11] Plant-based
medicines are widely used in developing countries and the
World Health Organisation is supporting research and development of their use. The WHO has urged Third-World
governments not to rely exclusively on Western-type medicine
but to aim for a synthesis '. . . between the best of modern with
the best of traditional medicine.'[11]

Practitioners of herbal medicine use the same methods to
diagnose ill-health as conventional doctors but herbalists
strongly believe that medicines derived from whole plants are
better balanced and less prone to harmful side-effects than
conventional drugs, which are concentrates of isolated plant
substances or artificial chemical compounds.[46] For instance, the
toxicity of digitalis is much reduced when patients are given the
whole plant. The initial side-effect of the plant is only nausea
whereas the isolated drug can produce heart arrhythmias. The
dandelion plant is a valuable diuretic and has the added
advantage that it contains three times the potassium content of
an ordinary green-leaved plant. Most chemical diuretics,
unfortunately, leach potassium from the body. The herb
ephedra's active ingredient is ephedrine, used to treat asthma.
On its own the drug increases blood pressure and heart rate but
the whole plant contains alkaloids that counter these side-effects
– a considerable advantage to patients. Unfortunately, whole
plant remedies do not offer the same financial inducements as
synthetic drugs which can be patented, giving pharmaceutical
firms monopoly rights on their sale.

Many alternative approaches are currently being mobilized in
projects such as the Cancer Help Centre in Bristol where
therapies are designed to revitalize the body's immune system

rather than concentrating on the tumour site itself.[47] Natural diets, vitamin supplements, exercise, group therapy and support, relaxation and counselling are all used to provide patients with the positive attitude and inner resources necessary to defeat the disease. Another technique in use at the Centre is biofeedback, which enables patients to harness the power of the mind to reinforce the body's natural defences. And at the Lemual Shattuck Hospital in Boston, Massachusetts, over 30 therapies are used to treat patients with chronic pain.[46]

To many, conventional medicine, heavily reliant on powerful, often toxic drugs, now seems a narrow approach to health, unable to cope with the diseases of civilization. All too often modern medicine actually perpetuates disease by ignoring the causes and simply treating the symptoms. The growing popularity of alternative medicine and the success of new methods such as visualization techniques, in which patients use their imagination to stimulate the body's natural defences against diseases like cancer,[48] show how much more medicine has to offer. By recognizing the importance of both mind and body, alternative therapies promise a more enlightened and far less mechanical view of disease and its treatment. In the years ahead, only the best of conventional medicine is likely to survive; the rest should not and hopefully will not.

1 Health Action International, Press Release and 'Problem Drugs' pack, 13 May, 1986
2 *SCRIP*, 7, 14 January, 1987
3 *New Internationalist*, November, 1986
4 *BMJ*, 496, 15 February, 1986
5 *Lancet*, 632, 14 September, 1974
6 G. Nuki, *BMJ*, 39-43, 2 July, 1983
7 MRC Annual Report, 1982-83
8 *SCRIP*, 27, 17 June, 1985
9 R. Rondell recorded in reference 18
10 M. Silverman, et al, *International Journal of Health Services*, 585-596, volume 12, 1982
11 D. Melrose, *Bitter Pills* (Oxfam, 1982)
12 J. S. Yudkin, *Lancet*, 1114, 14 November, 1981
13 *New Scientist*, 8, 2 May, 1985 and *Hansard* 50-51, 30 April, 1984
14 J. Collier and L. New, *Lancet*, 341-342, 11 February, 1984
15 *Hansard*, 354, 21 January, 1985
16 C. Medawar, *The Wrong Kind of Medicine* (Consumer Association and Hodder & Stoughton, 1984)

17 J. W. Dunne in reference 18
18 T. Smith, *BMJ*, 1255-1257, 8 November, 1980
19 D. Sanders, *The Struggle For Health* (Macmillan, 1985)
20 R. Blum, A. Herxheimer, C. Stenzl and J. Woodcock (Eds.) *Pharmaceuticals and Health Policy* (Croom Helm, 1981)
21 A. Melville and C. Johnson, *Cured to Death* (New English Library, 1983)
22 'The Opren Scandal', Panorama, BBC1, 17 January, 1983
23 See Chapter 3
24 M. D. Rawlins, *Lancet*, 276-278, 4 August, 1984
25 R. D. Mann, *Modern Drug Use* (MTP Press Ltd, 1984)
26 J. P. Griffin and G. E. Diggle, *British Journal of Clinical Pharmacology*, 453-463, volume 12, 1981
27 H. F. Steward in *Providing for the Health Services*, B. Black and G. P. Thomas (Eds.) (Croom Helm, 1978)
28 *SCRIP*, 20-21, 23 December, 1985
29 'Statistics of experiments on living animals', Home Office, 1986 (HMSO, 1987)
30 Reproduced in reference 1
31 W. H. Inman (Ed.) *Monitoring for Drug Safety* (MTP Press Ltd., 1980)
32 *BMJ*, 1397-1398, 24 November, 1984
33 *Technical Report Series of the World Health Organisation*, no. 615, 1977 reproduced in reference 11
34 H. W. Balme, *Lancet*, 293, 4 February, 1984
35 *The Guardian*, 9 March, 1983
36 *Lancet*, 531, 2 March, 1985
37 *Lancet*, 497, 2 March, 1985
38 F. Lesser, *New Scientist*, 10-11, 10 January, 1985
39 *BMJ*, 307-308, 30 July, 1983
40 D. T. Reilly, *BMJ*, 337-339, 30 July, 1983
41 *The Guardian*, 2 October, 1986
42 *Rheumatology in Practice*, 15, January, 1986
43 *New Scientist*, 19, 18 July, 1985
44 *Lancet*, 875-877, April 18, 1981
45 *New Scientist*, 464, 14 February, 1980 and Jing-Bo Jiang, et al, *Lancet*, 285-288, 7 August, 1982
46 T. Kaptchuk and M. Croucher, *The Healing Arts* (BBC Publications, 1986)
47 *Progress Without Pain*, video produced by the Lord Dowding Fund for Humane Research, 51 Harley Street, London W1
48 B. Inglis, *The Diseases of Civilization* (Paladin Books, Granada Publishing Ltd, 1983)
49 *Drug & Therapeutics Bulletin*, 23 March, 1987

PART 3

The Alternative

CHAPTER 5

The advances

'The history of medicine has shown that, whenever medicine has strayed from clinical observation, the result has been chaos, stagnation and disaster.'

(Professor A. P. Cawadias
Science, Medicine and History, 1953[1])

In 1846 Ignaz Phillipe Semmelweiss, then only 26 years of age, joined the obstetric staff at the Allgemeines Krankenhaus in Vienna. It was a crucial appointment, for in the months ahead Semmelweiss would show how the appalling mortality from puerperal, or child-bed, fever could be dramatically cut and the disease banished.

By the time Semmelweiss arrived, the first ward in the hospital had acquired such a bad reputation on account of its high death rate that expectant mothers, with tears in their eyes, begged not to be placed in it. He had found an expectant mother crying because she had been assigned to the students' ward – the first ward – rather than to the midwives' and this had come to be regarded as a death sentence.[2]

Semmelweiss soon noticed that students entered the first ward for their instruction in obstetrics, straight from the dissecting rooms, whilst in the second ward, with a much lower mortality, the work was done by midwives who had nothing to do with the dissecting and post-mortem rooms.

Then, in 1847 came the clue he needed. A colleague fell victim to blood poisoning caused by a wound inflicted on himself whilst carrying out a post-mortem.[2] And the symptoms, Semmelweiss observed, were similar to those of women who had died of puerperal sepsis. Convinced now that puerperal fever was due to an infection carried from the dissecting room on the hands of doctors and medical students, he issued strict orders that their hands should be thoroughly washed between each case they attended. His theory proved correct and the death rate promptly dropped from one in eight confinements to one in a hundred.[3]

Semmelweiss reached his conclusions by astute detective work based on clinical observation of his patients, and if his methods were not glamorous, they were certainly effective. But Semmelweiss could not point to a responsible germ, *merely* that his patients survived whilst others died and soon, after considerable hostility from his colleagues, he was forced to leave.

This terrible indictment of medical science – that results must be 'proven' in the laboratory before finally being accepted – encouraged the belief that only laboratory research held the key to health. Indeed, Louis Pasteur, the famous French chemist usually credited with the germ theory of disease, actually described laboratories[4] as 'temples of the future' and his countryman and fellow vivisector Claude Bernard referred to them[5] as 'the true sanctuary of medical science.'

Today, thanks to sustained propaganda, most people accept the myth that laboratory research, and animal experiments in particular, are the key weapons in the fight against disease. Yet history shows that it is really clinical investigation, together with intelligent application of chance discoveries, that have provided most of the really important medical advances.

It is to Hippocrates, the Father of Medicine, that we owe the birth of clinical medicine.[1] Up until the fifth century BC medicine had all too often relied on magical rites and witchcraft, but Hippocrates encouraged doctors to discover the physical causes of disease rather than the supernatural. He stressed the need to observe symptoms and accurately record the physical signs of illness and by so doing converted medicine into an art that must be studied and mastered by the slow process of trial and error.

Great attention was given to exact observation so that doctors would be able to profit by their previous experience and intervene earlier and more effectively on their patients' behalf.[6] Hippocrates placed far more reliance on general measures such as diet, rest and hydrotherapy, to stimulate the body's own natural healing powers, and therefore avoided poisonous drugs.[2] Clinical experiments were also carried out, for instance using test meals to investigate the patient's digestion.[6] But if Hippocrates taught that medicine should be learnt at the bedside, there would soon come a man who would change all that.

The physician who was destined to dominate medicine for many centuries and who wrote with such conviction and dogmatism that few dared to criticize him, was born at Pergamos in Asia Minor in the year AD131. Galen made his reputation in Rome but left shortly before an epidemic of plague. That he was later persuaded to return, '. . . has been widely considered', writes Brian Inglis in his *History of Medicine*, '. . . one of the greatest misfortunes that medicine has ever suffered.'[2] Unlike Hippocrates, Galen left no good accounts of clinical cases, being more interested in extolling his miracle cures.[7] Instead of objectively interpreting facts as observed at the bedside, he fitted everything he saw to the fashionable theories of the day, including that of the four humours. And if the facts did not fit the theory then the facts had to suffer.

As the founder of experimental physiology, Galen vivisected many animals and, because he unhesitatingly transferred his results to human beings, there were many mistakes.[7,8] Human anatomy and physiology, then, were based almost entirely on animals. Yet despite his many dissections and experiments on animals, Galen believed that the blood, in its transit through the body, passed through the heart by means of invisible pores in the intervening septum. Unfortunately this idea prevented doctors from having a real insight into blood circulation until the seventeenth century.[7]

Whilst Hippocrates had insisted on strict cleanliness during operations, expecting wounds to heal without infection,[9] Galen believed that the formation of pus was an essential part of the healing process, an error that greatly retarded the progress of surgery.[8,10] This is the horrible doctrine of 'laudable pus' that led to the idea that wounds should be deliberately irritated and contaminated.[11] Even in the nineteenth century surgeons would still talk of laudable pus. So perhaps it is not surprising that after the fall of the Roman empire hygienic measures lost their importance. Rome had at least taught doctors the importance of drainage and clean water supplies but in the Middle Ages towns were completely deficient in both of these. They had no drains and no clean water supply, the houses were ill-ventilated and filthy, the streets narrow and foul-smelling.[6] But what can be wrong with dirt when even surgeons do not insist on cleanliness?

Galen's dogmatic style, together with the Church's reluctance to allow dissection of human cadavers, meant that his errors became enshrined in medical teaching for nearly 14 centuries. Right up until the time of Vesalius, everything relating to anatomy, physiology and disease was referred back to Galen as a final authority from whom there could be no appeal. Few had the courage or the desire to embark on fresh clinical observations. And why should they if Galen had already discovered all they needed to know? As Garrison's *History of Medicine* puts it: 'After his death, European medicine remained at a dead level for nearly 14 centuries.'[7]

Then, after centuries of stagnation, came the Renaissance. Like several of the great artists, Leonardo da Vinci started to dissect the *human* body and to make drawings of its deeper structures. After carefully dissecting the heart and large blood vessels he discovered the descriptions of them given by Galen were gravely at fault.[6] By carrying out experiments on his human cadavers, Leonardo was able to show that the heart valves were so placed that they ensured the blood would always flow in one direction only and not be able to regurgitate back into the heart, so dispensing with the old ebb and flow theory. But the idea that blood percolated from one side of the heart to the other through pores in the intervening septum was so firmly fixed in everyone's mind that even Leonardo was unable to avoid it. As a result he failed to anticipate Harvey's discovery of blood circulation made a century later.

Of even greater significance was the meteoric rise of a young anatomist called Vesalius. As a student he grew tired of hearing the works of Galen and went to study anatomy at Padua, where a certain amount of dissection of the *human* body was being carried out. He made such a good impression that he was soon elected Professor of Anatomy and students flocked from all parts of Europe to hear his anatomical lectures. In 1543, after years of painstaking work, he published his great work *The Structure of the Human Body*.

Vesalius discovered that Galenic anatomy was based on animals and consequently much of what Galen had presented as human anatomy was mere imagination. Galen was held to blame not just for his faulty results but for using the wrong method.[12] Inevitably there was a backlash and a terrible storm broke over his head: for Vesalius had been guilty of doing what no man had hitherto dared to do – he had contradicted the great Galen. He was abused, derided and discredited and eventually forced to abandon his work.[6]

Almost 20 years before, the Swiss physician Paracelsus had similarly scandalized his colleagues after publicly burning the works of Galen and Avicenna. 'This', he announced, 'is the cause of the misery of this world, that your science is founded upon lies. You are not professors of the truth, but professors of falsehood.'[6]

However, Vesalius's work had not been in vain. He had imparted a new vigour to the study of anatomy, and human dissections were now being carried out in all of those centres where the Church did not expressly forbid them. In the UK an Anatomy Act was passed that gave the Barber-surgeons of London the right to dissect the bodies of four executed criminals each year.[6] For Galenism it was the beginning of the end: the foundation for the future teaching of medicine, particularly anatomy and surgery, had been laid.

Another crucial Renaissance figure was the great French surgeon Ambroise Paré who, inspired by the anatomical work of Vesalius, led surgery out of the Middle Ages.[11] Because his elementary schooling had been poor, he knew neither Latin nor Greek and so was debarred from entering the University of Paris. In fact, this supposed handicap was to prove a blessing in disguise, for Paré was spared the teachings of Galen and compelled instead to study Nature herself, as the great Hippocrates had done.[6]

Paré was primarily a military surgeon and his knowledge derived from observations made in the battlefield, which has always been a great teacher of surgery. Paré was the first to treat serious bleeding by finding the cut artery and tying a ligature round it instead of burning it.[11] He also abolished the use of boiling oil to cauterize gunshot wounds. The terrible cruelty of it horrified him and it chanced that in 1536, when attending a French force to relieve Turin, Paré ran out of oil. Instead he applied a bland dressing of egg-yolk, oil of roses and turpentine. In the morning his patients were comfortable and their wounds not inflamed but others, treated with the boiling oil, were feverish and in great pain from their festering wounds.[6,9]

Ambroise Paré had taken surgery as far as it could reasonably expect to go and fresh advances would have to wait until the nineteenth century with the discovery of anaesthetics and the rebirth of hygienic principles.

The next important advance came in 1628 when William Harvey, formerly a student at the Padua school of anatomy, published his treatise on the circulation of the blood. It is widely believed that Harvey was the first to discover blood circulation but in 2650 BC Chinese Emperor and scientist Hwang Ti wrote:

> 'All the blood in the body is under the control of the heart. The blood current flows continuously in a circle, and never stops.'[13]

Harvey confirmed Leonardo's finding that the heart valves are so arranged as to make it possible for the blood to travel in one direction only. But his most famous experiment involves binding the arm above the elbow so that the veins stand out, with their valves seen as knots or swellings.[11] By compressing a vein with a finger, the blood is milked upwards to the next valve. The vein between the finger and the valve remains empty until the finger is released and then the vein fills from below. This is repeated until so much blood has passed along the vein, always in the same direction, that '. . . you will be completely convinced, by the speed of the blood's movement, of the fact that the blood circulates.'[14] He also carried out autopsies to investigate the anatomy of the heart:

> 'From the structure of the heart it is clear that the blood is constantly carried through the lungs into the aorta as by two clacks of a water bellows to raise water. By application of a bandage to the arm it is clear that there is a transit of blood from arteries to veins wherefore the beat of the heart produces a perpetual circular motion of the blood.'[15]

In view of the simple experiments he carried out on himself and on human cadavers, it seems strange he should resort to experiments on animals yet he claims[6] to have used no less than 80 different species! Perhaps he anticipated a hostile response in challenging Galen's entrenched ideas and wanted to show he had conducted sufficient experiments to prove his point, which may explain his ten-year delay in publication.[2]

Later on, in response to opposition from Riolanus, Dean of the College of Paris, Harvey wrote to a friend in Hamburg, explaining how the simple experiments he had carried out on the body of a hanged criminal would '. . . easily put an end to all Riolanus's altercations on the matter.' In the presence of several colleagues, Harvey had forced water first into the right side of the heart and then the left, in each case watching the direction and course of the fluid.[16]

The discovery of blood circulation, then, was not dependent on animal experiments but careful *human* observations. According to Lawson Tait, one of the greatest surgeons of the nineteenth century:

> 'That he [Harvey] made any solid contribution to the facts of the case by vivisection is conclusively disproved, and this was practically admitted before the Commission by such good authorities as Dr Acland and Dr Lauder Brunton. The circulation was not proved till Malpighi used the microscope and though in that observation he used a vivisectional experiment his proceeding was wholly unnecessary, for he could have better and more easily used the web of the frog's foot than its lung.
>
> 'It is, moreover, perfectly clear that were it encumbent on any one to prove the circulation of the blood as a new theme, it could not be done by any vivisectional process but could, at once, be satisfactorily established by a dead body and an injecting syringe.'[17]

Through the long period of Galenic darkness, the teachings of Hippocrates had been largely forgotten. Now, thanks to London physician Thomas Sydenham, they were to be revived and the general level of medical practice raised. Sydenham modelled himself on Hippocrates and insisted that doctors must learn about disease at the bedside and rely on clinical observation rather than base their treatment on a general theory of disease as Galen had done.[6] He introduced the use of iron in cases of anaemia, popularized the treatment of malaria with cinchona bark from Peru and treated syphilis with mercury.[11] But by far his greatest contribution was his insistence on a return to Hippocratic principles at a time when medical men were once again losing themselves in medical theories and philosophical speculations. If Vesalius's work forms the basis of modern anatomy and surgery, being founded on careful dissection and observation of human cadavers, then it is to Thomas Sydenham that we owe the return to the long forgotten principles of clinical investigation – the only truly valid approach to medical research.

Generally though the post-Renaissance period was not a time of great *therapeutic* advance. Harvey's research had no immediate practical application and it would be some time before the more rational approaches established by Vesalius and

Sydenham actually bore fruit. The sick enjoyed few new benefits and they ceased to have many of the advantages of older times – devoted Christian care, sanitation, hygiene, diet, rest and quiet.[2] The eighteenth century too was not a period that brought many decisive new advances in treatment but, nevertheless, certain therapeutic truths known to medicine in earlier times were rediscovered.

In 1718 Lady Wortley Montague, wife of the British Ambassador to Turkey, introduced inoculation against smallpox into this country.[3] Small amounts of material from the pustules of those suffering a mild form of the disease were administered nasally or inoculated into those seeking protection, immunity being conferred against dangerous attacks. But inoculation had been practised in India since ancient times and in China since 1063. It had even been realized that risks could be diminished by leaving fluid from the pustules in the open air, thereby reducing virulence. Yet it is often thought that Edward Jenner discovered vaccination against smallpox and with it the whole principle of immunity, as if the Indians and Chinese never existed! Even so, Jenner's 'discovery' was another example of clinical investigation, albeit rather primitive.

A young farm girl who visited him as a patient boasted that she was not afraid of catching the disease, because she had already had cowpox.[6] Milkmaids commonly caught the disease from lesions on the udders of cows and, on further investigation, Jenner learned how farmers strongly believed that an attack of cowpox made people safe against smallpox. A hired hand who had had cowpox had a better chance of a job than one who had not and people who had contracted the milder disease were employed as nurses in an epidemic of smallpox. Jenner himself tells the story of a matron who had accidentally been infected with cowpox through handling infected milk pails but when she attended a relative who died of smallpox, did not herself get the disease.[18]

Jenner decided to put the theory to the test and his chance came in 1796 when, during an outbreak of cowpox on a neighbouring farm, Sarah Nelmes, a dairymaid, became infected. She had stuck her finger with a thorn and had a tiny scratch through which the disease matter entered. Jenner collected fluid from the pustule on Sarah's hand and vaccinated James Phipps, an eight-year-old boy. A few weeks later he injected the child with fresh smallpox fluid taken from a blister.

Fortunately the child remained well, presumably now immune. Jenner then inoculated smallpox matter into ten other people who had already had cowpox and they too resisted the disease. Apart from his experiment with James Phipps who would be unable to give any kind of informed consent, Jenner's work was based entirely on clinical observation, and in 1798 he published his results.

Today, almost 200 years after Jenner's original trials, medicine is obsessed with vaccines: scientists are even trying to develop one against tooth decay as if oral hygiene is no longer sufficient.[19] Yet their overall contribution has not been great, except for their manufacturers. Some, such as Pasteur's rabies vaccine, were positively hazardous,[20] some had little demonstrable effect and nearly all were introduced when the disease itself was already in retreat. And compared with improved nutrition and environmental factors, their impact was negligible.[21]

During the eighteenth century, too, attention was once again drawn to the need for a balanced diet by James Lind's dramatic treatment of scurvy. In 1740 Lord Anson took six ships on a world cruise but barely managed to struggle back to port again with a fleet of skeleton crews. Scurvy had claimed the lives of some 1,200 of his men.[6] Yet the cure had been known for centuries. In 1535 Jacques Cartier set sail with 110 men to explore the coast of Newfoundland and the River St Lawrence. After six weeks only ten of his crew remained unaffected. Then Cartier learned from a native '. . . that the juice and sappe of the leaves of a certain tree' cured this complaint from which the native himself had previously suffered. The ravaged crew were given the vegetable extract and to Cartier's great relief, quickly recovered.

Other accounts too describe the use of fruit and vegetables to prevent scurvy and, in 1564, on a return journey from Spain, Dutch sailors suffered a severe attack of scurvy and cured themselves by eating some of the oranges and lemons that made up the main bulk of their own cargo. In 1600 Captain Lancaster, commanding HMS Dragon, ordered a daily issue of three spoonsful of lemon juice.[3] When he dropped anchor in Table Bay, South Africa, his crew were in perfect health and able to row ashore the sailors from three other ships who had made the same voyage but who, not having the same treatment, were incapable of wielding an oar. And in 1617 John Woodall wrote that a daily dose of lemon, orange or lime juice was necessary to avoid scurvy.[3]

But, as so often happens in medicine, valuable knowledge is forgotten or ignored and has to be rediscovered at a later date. So it is that James Lind, over 200 years after Cartier's fateful voyage, is usually credited with the discovery that scurvy is a deficiency disease that can be cured or prevented by giving fresh fruit or vegetables.

Lind had become all too familiar with scurvy during his service as a naval surgeon and he was also aware of Cartier's discovery. So in 1747, on board HMS Salisbury, he decided to carry out a clinical trial in which some of his patients with scurvy were treated with oranges and lemons whilst the others were given nondietary remedies. The test worked beautifully but it was not until 1795, the year after Lind's death, that the Admiralty were persuaded to make the use of lemon juice compulsory in the Navy.[3,6]

One hundred and sixty years *after* Lind's clinical study, in 1907, the Norwegians Holst and Frölich produced scurvy in guinea pigs by cutting out cabbage leaves from their diet and feeding them on nothing but grain and water.[3] (Even so the choice of species was a lucky one[22] for many animals do not need to have vitamin C added to their diet).

Much the same happened with beri beri. As early as the tenth century Chinese writers were blaming rice as the cause of this deficiency disease and in 1882 Admiral Takaki effectively eradicated beri beri from the Japanese Navy by insisting on a more liberal and better balanced diet, which, until then, had consisted mainly of polished rice, that is, rice without the husk.[3] Fifteen years later, in 1897, Eijkman and Grijns succeeded in giving the disease to pigeons by feeding them solely on polished rice.[3] The need for other important vitamins, such as A and D, nicotinic acid and B_{12} were also first discovered in people and only later confirmed in laboratory animals.[22,23]

But all this is a prelude to the astonishing revival of medicine during the nineteenth century. Two key advances – the return of hygiene and the discovery of anaesthetics – enabled surgery to advance rapidly, whilst social and humanitarian reformers like Chadwick in Britain and Shattuck in America initiated vital public health measures that would eventually lead to the virtual eradication of the infectious epidemics. Over the next 150 years the dramatic fall in deaths could be directly traced to the improvements in nutrition, living and working conditions, hygiene and sanitation (see also Chapter 1).

Until the mid nineteenth century, patients preparing for an operation must have felt like condemned criminals preparing for execution and there was little hope for improvement unless surgery could exorcise two ever-present spectres: the fear of pain during the operation and the risk of infection afterwards. The most valued surgeons, therefore, were the quickest and it is said that William Cheselden (1688-1752) was able to perform a lithotomy (removal of a stone from the bladder), in a minute and that on one occasion he managed it in only 54 seconds![6]

That certain plants could be used as pain killers had been known for a very long time: opium was in use as early as 3,000 BC and the narcotic effects of Indian hemp were known in ancient times in the East. Indeed, surgeons were using it in China before operations in the second century AD. Yet by the eighteenth century, Western patients about to be operated upon would be lucky to receive anything to kill the pain. Then in 1800 Humphrey Davy inhaled nitrous oxide or 'laughing gas' and noted that it eased the pain of an inflamed gum and suggested its use in surgical operations.[6] But inhaling laughing gas was regarded as a kind of parlour game and the proposal was ignored. The *Encyclopaedia of Medical History* records how '. . . there was no systematic search for effective anaesthesia, and when one was finally discovered, its discovery was fortuitous.'[24] And so it was.

By the 1840s, laughing gas parties and 'ether frolics' were popular entertainments, especially among medical students, and it was his experience whilst inhaling ether that first prompted Dr Crawford Long of Jefferson, Georgia, to suggest its use during surgical operations:

> 'On numerous occasions I inhaled the ether for its exhilarating qualities, and would frequently at some short time discover bruises or painful spots on my person which I had no recollection of causing, and which I felt satisfied were received while under the influence of ether. I noticed my friends while etherized received falls and blows, which I believed sufficient to cause pain on a person not in a state of anaesthesia, and on questioning they uniformly assured me that they did not feel the least pain from these accidents.
> 'Observing these facts I was led to believe that anaesthesia was produced by the inhalation of ether, and that its use would be applicable in surgical operations.'[2]

In 1842 Long put his theory to the test and successfully removed a cystic tumour from the neck of a young man called James Venables.[25] There was not the slightest pain and Long subsequently used ether in other cases.

Two years later, Horace Wells, a dentist from Hartford Connecticut, began to use nitrous oxide in dentistry, communicating his results to his friend and former partner William Morton who in 1846 removed a deep-rooted tooth from a patient, Eban H. Frost, using ether as anaesthetic.[25] Morton then visited Dr John Warren of the Massachusetts General Hospital and persuaded him to give ether a trial in surgical practice. The operation took place on October 16, 1846 when Warren successfully removed a tumour from the neck of a young man called Gilbert Abbott. As the patient recovered from the anaesthetic, Warren announced, 'Gentlemen, this is no humbug.'[2]

The news of ether spread rapidly and prompted Edinburgh gynaecologist James Simpson to investigate other vapours that might also have anaesthetic properties. He had not found ether entirely satisfactory in gynaecological work but as his practice was very large, the only time available for testing new vapours was normally at the end of a long day's work. Professor Miller, one of Simpson's colleagues, describes how the anaesthetic properties of chloroform were discovered:

'Late one evening – it was the 4th of November 1847 – on returning home after a very weary day's work, Dr Simpson, with his two friends and assistants, Drs Keith and Duncan, sat down to their somewhat hazardous work in Dr Simpson's dining room. Having inhaled several substances, but without much effect, it occurred to Dr Simpson to try a ponderous material which he had formerly set aside on a lumber table, and which on account of its great weight, he had hitherto regarded as of no likelihood whatever. *That* happened to be a small bottle of chloroform. It was searched for and recovered from beneath a heap of waste paper. And with each tumbler newly charged the inhalers resumed their vocation. Immediately an unwanted hilarity seized the party; they became bright-eyed, very happy and very loquacious – expatiating on the delicious aroma of the new fluid . . . a moment more and then all was quiet – and then *crash*. The inhaling party slipped off their chairs and flopped onto the floor unconscious.'[6]

Within a fortnight Dr Simpson had administered chloroform to at least 50 of his patients with excellent results.[6] In the early years of the twentieth century, the second Royal Commission on Vivisection would officially confirm that 'The discovery of anaesthetics owes nothing to experiments on animals.'

The appearance of ether and chloroform inevitably cut short the promising work of James Esdaile who had performed hundreds of serious operations, including amputations, and thousands of minor ones, under hypnotic trance. They had proceeded quite painlessly and with a negligible fatality rate, but when he returned to Britain the editors of the medical journals refused to publish his findings: 'They will not admit', he wrote to a colleague, 'or permit you even to hear of, such indisputable facts, through fear of the consequences'[2] – the disapproval of the medical establishment. Today hypnosis is being used in dental surgery as a safe and acceptable alternative.[26]

For patients, the discovery of surgical anaesthetics must have seemed like a miracle, yet without a return to the old hygienic principles, they would still be lucky to survive the ordeal. So great was the risk of postoperative infection that Sir James Simpson, the discoverer of chloroform, once said: 'the man laid on an operating table in one of our surgical hospitals is exposed to more chances of death than was the English soldier on the field of Waterloo.'[6] At that time an operation was almost invariably followed by fever and pain. A large number died of hospital gangrene and the surgical profession spoke of 'laudable pus' because its appearance was considered necessary for the healing of wounds. Yet it had not always been so.

Indian surgeons in the fourth millenium BC put great emphasis on strict cleanliness, with regulations about washing hands and nails – even speaking was prohibited during an operation, in case the bystander's breath should contaminate the wound.[2] According to Sir Cecil Wakely,[9] consulting surgeon to King's College Hospital, Greek surgeons, following the teaching of Hippocrates in the fifth century BC, filtered or boiled the water used in washing wounds and '... their rules of cleanliness could not be bettered.' It was expected that wounds should heal by first intention, that is without infection. But gradually the need for cleanliness in assisting nature's great healing powers disappeared and surgery declined with it.

The surgeon usually credited with the development of surgical hygiene is Joseph Lister who in 1867 introduced the idea of antiseptic surgery. When Lister read of Pasteur's work on bacteria he immediately associated them with the infections he had so often encountered during surgery. Lister's method was to use a spray of carbolic acid in an attempt to slay all the surrounding germs. In fact it was not necessary to massacre all the microbes in the vicinity: all that was needed, as practised in ancient times, was strict hygiene. Fortunately a handful of surgeons opposed Lister's use of antiseptics on the grounds that they damaged the surrounding tissues and insisted instead on strict adherence to cleanliness or what came to be known as the *aseptic* technique.[2]

One of Lawson Tait's biographers, Harvey Flack, affirms that, '. . . for all his brief life Tait denied the theory of antisepsis'[27] relying instead on meticulous cleanliness to achieve his results. And another famous surgeon of his times, John Harvey Kellog, claimed that Tait was '. . . really the father of surgical asepsis.'[27] Another opponent was abdominal surgeon Dr Granville Bantock, also a medical antivivisectionist, who argued that all that was really necessary was to exclude dirt and keep wounds scrupulously clean:

> 'What is this but the doctrine of cleanliness, which I have advocated for so many years? Thus it will be seen that it only required that Lord Lister should have taken one more step to fall in line with me.'[28]

Eventually the aseptic technique won the day and even Lister himself acknowledged that:

> 'As regards the spray, I feel ashamed that I should ever have recommended it for the purpose of destroying microbes in the air.'[29]

Yet, by 1847, 20 years before Lister's 'discovery', there was already a mass of evidence in favour of asepsis,[3] apart from the surgeons of antiquity. On the basis of their clinical observations, several doctors had shown that puerperal, or childbed fever could be banished by the introduction of hygienic (or aseptic) measures. The link with surgical practice is clear because in the process of having a baby the whole inside of the womb is, in effect, wounded and therefore particularly susceptible to infection.

In 1795 Alexander Gordon published his *Treatise on the Epidemic Puerperal Fever of Aberdeen* and demonstrated the contagious nature of the disease. He advised thorough disinfection of the hands and clothes of doctors and midwives. Then in 1843 Boston gynaecologist Oliver Wendell Holmes wrote *The Contagiousness of Puerperal Fever*, showing that childbed fever was an infectious disease carried from one patient to another on the hands of the attendant or midwife. And in 1846/7 came the most famous rediscovery of all when Semmelweiss once again revealed the contagious nature of the disease and ordered thorough washing of hands, instruments and dressings, together with the isolation of infected women. But if Gordon and Holmes were largely ignored, all that Semmelweiss managed to achieve by publicising the successful results of these simple measures was to infuriate his colleagues and he was promptly sacked.[2] Undeterred he made a fresh start at the St Rochus Hospital in Budapest where he again achieved dramatically successful results. It was here in 1861 that he wrote his work *The Etiology, Concept and Prophylaxis of Childbed Fever*. Soon, however, the clinical scientists would be proved correct when the responsible microbes were isolated from patients and detected under the microscope.[30]

Now that the two great barriers to surgical progress – the fear of pain and infection – had been lifted, surgery advanced rapidly and, as clinical experience increased, techniques improved and operations, hitherto unthinkable, became possible (see Chapter 3, page 79). And in 1895 when physics professor Wilhelm Röntgen accidentally discovered X-rays, another valuable technique that owed nothing to animal experiments was added to the surgeon's armoury.[6] Röntgen was busy passing electrical discharges through a partially evacuated glass tube when he discovered, quite by chance, that highly penetrative but invisible rays were being emitted from the tube. Röntgen called them X-rays and found that wood, glass, sheet metal and human flesh, but not bone (he had put his own hand in the path of the rays), were all transparent to the rays.

The use of cocaine as a local anaesthetic and curare as a muscle relaxant provided surgeons with additional aids, both remedies coming from the Incas of Peru.[6] A further boost came in the early years of the twentieth century with the advent of safe blood transfusions. But this time it could be argued that animal experiments had actually retarded progress.

In 1666, following Harvey's discovery of blood circulation, the Cornishman Richard Lower transferred blood from the artery of one dog into the vein of another and a year later the

French physician Jean Denis transfused lamb's blood into a boy. But a second patient died and his widow promptly brought a lawsuit against the Professor.[6] Unfortunately for Denis, and his patients, human and animal blood are totally incompatible, which makes transfusions with blood from animals not only useless but highly dangerous.[6] And when more patients died no further attempts were made for more than a century.[24,31]

In France blood transfusions were considered so hazardous that the Paris Faculty of Medicine actually sought their prohibition through an Act of Parliament.[31] Eventually, in the early years of the nineteenth century, it was realized that transfusion must rely on blood from *human* donors.[24] Blundell had used animal experiments to prove that blood should only be transferred between the same species but, ironically, if vivisection had never been allowed, the original dangers would never have arisen! Even so, the method only became safe when Karl Landsteiner discovered the main blood groups in 1900.

If a patient is given blood against which he has antibodies, his body will destroy the transfused blood, setting up a severe reaction sometimes leading to kidney failure and death. Landsteiner was able to explain these observations by test tube experiments with human blood: by mixing different blood samples, he found that the cells all clumped together, and this resulted from the presence of two antigens that he labelled A and B.[24]

Landsteiner discovered that all human blood could be classified according to the presence or absence of these antigens. Using O for the absence of antigens, he found that A and O accounted for 41.8 per cent and 46.4 per cent of all people respectively; B appeared in 8.6 per cent and AB in just 3 per cent.

The ABO system made it possible to match donor and recipient blood types and, when test tube experiments and clinical studies showed that sodium citrate could prevent clotting,[32] the two main problems in the development of safe blood transfusions had been overcome. Even the discovery of the other important blood group – the rhesus factor – first came about through clinical observation, the name being derived when Landsteiner subsequently found the same blood group in rhesus monkeys.[33]

Another advance of considerable importance to medicine was bacteriology. Although it is the French chemist Louis Pasteur who is usually given the credit, he was far from being the first to advance a germ theory of disease. The idea is formulated in a treatise *De Re Rustica* published in the first century AD, in which Tarentius Rusticus writes: 'If there are any marshy places, little animals multiply which the eye cannot discern, but which enter the body with the breath through the mouth and nose and cause grave diseases.'[6]

Again, in the sixteenth century the idea was revived by Hieronymus Fracastorius in his work *De Contagione*,[2] where he divided infectious diseases into three categories: direct contagion through touch, indirect contagion through, say, sleeping on the same sheets that an infected person had slept on and airborne transmission on the same basis as pollen – the disease agents, Fracastorius believed, being tiny living particles, '. . . which multiply rapidly and propagate their like.'[6]

It was not until 1683, however, when Leeuwenhoek, with the aid of an efficient home-made microscope, actually discovered micro-organisms that a firm basis was provided for the germ theory of disease. Because Leeuwenhoek possessed great skill in grinding lenses, his microscopes were greatly superior to any previously made and with them he examined films of moisture taken from between his teeth.[6] To his amazement he discovered minute living creatures that he later called 'animalculi'.

After Leeuwenhoek had shown the way, other microscopists discovered 'animalculi' in the urine, faeces, blood and tissues of disease victims[3] and the solid basis for the germ theory had been laid. Discoveries now came thick and fast: in 1820 Ehrenberg made the first cultures of microbes using the cut halves of apples as his medium; Agostino Bassi proved that a fungus was the cause of muscardine, a disease of silkworms, and published his findings in 1836; micro-organisms associated with several skin diseases were isolated and identified, including thrush (1842) and ringworm (1846); Gros found the first parasitic amoeba in man in 1849, and so on.

Then Pasteur entered the field. First he established the role of micro-organisms in the fermentation of alcohol and discovered the agents that cause sourness in milk (1857), in butter (1861) and in wine (1863). Continuing his researches in an attempt to discover the origin of these bacteria, he came to the conclusion that they were carried in dust particles in the air and proved that a boiled liquid would not ferment when exposed to filtered or pure mountain air.[3] As a result of these

experiments Pasteur once and for all banished the idea of
spontaneous generation: life must come from life. A century
before, Spallanzani had also shown that fermentation only
occurred when air could reach a previously sterilized
substance.[24]

If microbes make milk, butter, wine, beer and silkworms sick,
why not people too? The belief that illnesses were caused by
micro-organisms had long been in Pasteur's mind and in the
1850s, in a paper dealing with fermentation, he had written:
'Everything indicates that contagious diseases owe their
existence to similar causes.'[6] So far, apart from his observations
on silkworms, Pasteur's research had not involved animals, but, as
we have seen, his work on fermentation and the idea that bacteria
were carried through the air had done enough to convince Lister
that the grave risks of infection following a surgical operation were
caused by microbes invading the wounds.[34] He therefore set out to
destroy all the surrounding bacteria and in 1867 introduced his
antiseptic method of surgery.

During the 1870s Pasteur turned to animal experiments in
order to 'prove' what by now seems to have become *his* germ
theory.[34] He soon had a rival and in the years ahead both
Pasteur in France and a young German doctor named Robert
Koch, were busy inoculating disease after disease into their
animal victims. Koch even laid down a set of rules for
establishing final proof that a particular germ caused the disease
in question. They are called Koch's postulates:[6,35]

1) a specific microbe should always be present in every case of
   the disease and not in other diseases or in health
2) the organism should be isolated and grown in pure culture
3) and when inoculated into laboratory animals, should
   reproduce the same disease.

Ironically, Koch's own work was to prove that animal
experiments were not only unnecessary, they could be
dangerously misleading. When a causal organism is injected into
animals, it often gives rise either to no disease at all or one
bearing no clinical resemblance to the original malady. For
instance, when the pneumococcus, isolated from a typical case
of pneumonia, is injected into rabbits it does not produce
pneumonia but a general septicaemia instead. 'Thus', writes the
Lancet, 'We cannot rely on Koch's postulates as a decisive test of
a causal organism.'[35] Koch was to admit as much himself after
his investigation of cholera. At the head of the German Cholera

Commission, Koch visited India and took 50 white mice with him from Berlin. In a report for the *British Medical Journal* of 1884, Koch describes how he,

> '. . . made all kinds of experiments on them . . . Although these experiments were constantly repeated with material from fresh cholera cases, our mice remained healthy. We then made experiments on monkeys, cats, poultry, dogs and various other animals that we were able to get hold of, but we were never able to arrive at anything in animals similar to the cholera process.'[36]

Koch was forced to rely on clinical investigation, seeking the guilty microbe by means of the microscope in cases of human cholera. In 1883 he discovered the responsible organism – the 'comma bacillus', so called because it is shaped like a comma – and by careful observation identified its transmission by drinking water, food and clothing. Clinical studies then, if given the opportunity, could be just as convincing if not more so than experiments on animals, and Koch concluded:

> 'Should it prove possible later on to produce anything similar to cholera in animals, that would not, for me, prove anything more than the facts which we now have before us. Besides, we know of other diseases which cannot be transferred to animals, eg, leprosy, and yet we must admit, from all we know of leprosy-bacilli, that they are the cause of the disease.'[36]

'We must be satisfied', Koch argued, 'that we verify the constant presence of a particular kind of bacteria in the disease in question, and the absence of the same bacteria in other diseases.'[36] Effectively then, the germ theory of disease is proved not by experiments on animals but by careful clinical investigation. And soon Koch would provide an even better reason for avoiding animal tests.

In 1882, Koch had isolated the tubercule bacillus from patients dying of tuberculosis and, on this occasion at least, many of his animals – mice, guinea pigs and monkeys but not frogs or turtles – also succumbed when inoculated.[18] Koch must have been delighted, although later on he may have wished his animal tests had proved negative as in the case of cholera. For it was through experiments with infected animals that Koch eventually produced his highly acclaimed cure for TB: tuberculin, made from attenuated bacilli. The impact was enormous. From all over the world TB sufferers set out for

Berlin in the hope that tuberculin would save them. Unfortunately for Koch and his patients TB takes a different form in animals and gradually the realization grew that it was a failure. Worse still, in patients who had previously been infected, tuberculin caused the disease to flare up again and the tests ended in disaster.[37]

Inevitably, perhaps, the germ theory led to the belief that all that was necessary to combat disease was to massacre the responsible microbes, as, indeed, Lister had proposed. Or the bacteria or viruses could be isolated, their virulence diminished and a vaccine prepared, as earlier generations had done against smallpox. Already, after Koch's discovery of the tubercule bacillus in 1882, attention was being focussed entirely on the microbe itself,[3] despite strong suspicions that the disease was linked to poverty and poor living and working conditions. Laboratory science, it seemed, held the key to health and surely it was only a question of time before disease would be completely banished. It all seemed so simple. It is true that bacteriology not only enabled doctors to identify infectious diseases more accurately, it also gave them the opportunity to search for potentially effective drugs. The germs could be examined in the test tube and experiments carried out *in vitro* or, by now more fashionably, in laboratory animals. This led to the sulpha drugs that were introduced in 1935, six years before the arrival of penicillin for internal use.

It soon came to be realized, though, that germs did not actually cause the disease – they were simply the *agents*. Patients only became ill if their resistance had been lowered – for instance through stress or malnutrition – and it was this increased susceptibility that represented the real cause of disease. Germs, then, usually become dangerous only when the body could not cope with them. This explained why many organisms, formerly thought to *cause* the disease, were frequently found in healthy people who suffered none of the usual clinical symptoms. Max von Pettenkofer, the great public health expert who brought down Munich's typhoid mortality by his insistence on clean water, had proved as much in 1892 by the most dramatic experiment carried out on himself. He obtained what in ordinary circumstances would be regarded as a lethal dose of cholera germs, which he drank with no adverse affects other than mild diarrhoea.[2] Encouraged by his results, Pettenkofer's staff took the same dose and were also unharmed by it. According to this idea, writes medical historian Brian Inglis, '. . . microbes are more akin to looters in a city where law and order have broken down. They may prompt disciplinary measures, but there should not be any misguided notion that arresting or shooting them will save the city; that can only be accomplished by restoring law and order – or, in the case of medicine, by reviving the life force',[38] the body's natural resistance.

Today we know, as the social reformers had already proved over a hundred years before, that poverty, deprivation and unsanitary conditions are major causes of disease even within affluent Western societies, where the death rate for TB is ten times higher in social class V than for professional workers in social class I.[39] In poorer countries mortality from whooping cough is 300 times more common,[40] confirming the terrible effects of malnutrition and overcrowding. Infectious diseases can be combated, therefore, not only by avoiding contact with the germs – through improved hygiene and sanitation – but also by strengthening host resistance through better diet and living and working conditions.

Luckily it did not need Pasteur's germ theory to convince Bentham, Chadwick and Shattuck of the desperate need for reform. Their efforts were initiated well before the Frenchman's ideas had been presented[41] and were based entirely on careful observation of the surrounding conditions and their own common sense. They may never have seen a germ but it is to the social and humanitarian reformers that we owe the tremendous fall in deaths over the next 150 years. Even Pasteur's biographer René Dubois, himself a microbiologist, although accepting that bacteriology had proved valuable in improving diagnosis and providing a rational basis for treatment, had to acknowledge that the decline in mortality caused by infections,

> 'began almost a century ago and has continued ever since at a fairly constant rate irrespective of the use of any specific therapy. The effect of antibacterial drugs is but a ripple on the wave which has been wearing down the mortality caused by infection in our communities.[42]

Over the past 200 years few would doubt that the most crucial medical advances – both in terms of saving lives and alleviating suffering – have been the enormous improvements in public health and the development of surgery, made possible by the return of hygienic principles and the discovery of anaesthetics. These and the many other advances we have discussed have come about either through clinical investigation of patients or by intelligent application of chance discoveries. Even the germ theory and the birth of bacteriology did not need animal experiments but clinical observation, together with the skilled use of the microscope and the development of culture techniques.

With such a record of achievement one might naturally expect clinical investigation to continue to form the basis of modern medical research, particularly as animal models of human disease could give such disastrous results, as Koch had already proved. But during the nineteenth century, thanks essentially to just a handful of men, vivisection was transformed from an occasional method into the scientific fashion we know today.

Claude Bernard became Professor of Experimental Physiology at the College de France in 1855 and carried out such horrendous experiments on animals that even his wife Marie could no longer stand it. Marie was very fond of animals but to make matters worse Bernard sometimes brought his mutilated victims home with him to continue his observations. On one occasion he brought home a dog with an open wound in its side, that was suffering from diarrhoea and had pus running from its nostrils. Such cruelties sickened Marie and after many years of dissention they finally separated.[43]

Some idea of Bernard's state of mind comes from his description of the physiologist: '. . . he no longer hears the cry of animals, he no longer sees the blood that flows, he sees only his idea and perceives only organisms concealing problems which he intends to solve.'[5]

Bernard first sought fame as a playwright but, when advised to give up writing by a Sorbonne professor who had read his unpublished works, he took up medicine as an alternative.[8] Here again, as his own pupil Paul Bert subsequently had to acknowledge, Bernard was '. . . anything but a brilliant pupil'[44] and turned instead to physiology and animal experiments. Like Pasteur, he cherished the glamorous image of laboratory science and considered it far more important than clinical work with patients:

'I consider hospitals only as the entrance to scientific medicine; they are the first field of observation which a physician enters; but the true sanctuary of medical science is a laboratory; only there will he seek explanations of life in the normal and pathological states by means of experimental analysis.

'In leaving the hospital, a physician . . . must go into his laboratory; and there, by experiments on animals, he will account for what he has observed in his patients, whether about the action of drugs or about the origin of morbid lesions in organs and tissues. There, in a word, he will achieve true medical science.'[5]

Bernard had been a pupil of Magendie, himself one of the most callous vivisectors of the century as evidenced by an eye-witness account of one of his demonstrations, reported in the *BMJ* by Dr Latour:

'I remember once, among other instances, the case of a poor dog, the roots of whose spinal nerves he was about to expose. Twice did the dog, all bloody and mutilated, escape from his implacable knife; and twice did I see him put his forepaws round Magendie's neck and lick his face. I confess – laugh vivisectors if you please – that I could no longer bear this sight.'[45]

According to Guthrie's *History of Medicine*, 'Magendie seemed . . . to substitute experiment for thought, thrusting his knife here and there to see what would come of it, and prodding in all directions in the hope of finding some new truth.'[8]

Bernard considered Magendie to be the modern founder of experimental physiology but, in fact, it was Bernard, more than any other single man in the nineteenth century, who was responsible for establishing the vivisection method. He popularized the artificial production of disease by chemical and physical means[46] and set out vivisection's cruel and twisted logic, that '. . . we can save living beings from death only after sacrificing others.'[5] In 1865 he published his *Introduction to the Study of Experimental Medicine* in order to convince physicians that laboratory instruction in physiology was just what they needed: 'Every physician should have a physiological laboratory; and this work is especially intended to give physicians rules and principles of experimentation to guide their study of ex-perimental medicine.' And so successful was Bernard's message that both the public and a substantial proportion of the medical profession were soon convinced that experiments on animals were, and still are, the key weapon in the fight against disease. Yet according to Dr George Hoggan who worked under Bernard:

> 'We sacrificed daily from one to three dogs, besides rabbits and other animals, and after four years' experience I am of the opinion that not one of these experiments on animals was justified or necessary. The idea of the good of humanity was simply out of the question, and would be laughed at, the great aim being to keep up with, or get ahead of, one's contemporaries in science, even at the price of an incalculable amount of torture needlessly and iniquitously inflicted on the poor animals.'[47]

But, apart from the cruelty, Bernard's most disastrous legacy was his belief, reproduced in his *Introduction*, that animal experiments are directly applicable to humans: 'Experiments on animals, with deleterious substances or in harmful circumstances, are very useful and *entirely conclusive* for the toxicology and hygiene of man. Investigations of medicinal or of toxic substances also are *wholly applicable* to man from the therapeutic point of view . . .'[5] (emphasis added)

By their efforts Magendie, Claude Bernard and Louis Pasteur, through his attempts to develop vaccines against rabies and anthrax, the disease of cattle and sheep, helped earn France the reputation of being the country of vivisection. The now fashionable idea of developing animal models of human disease was further popularized by Robert Koch's insistence that the cause of infections must be confirmed by reproducing the illness in laboratory animals. Despite all of Galen's errors, they succeeded in turning animal experiments into an everyday practice, thus defying common sense. And because work with human patients requires so much more skill and patience in avoiding unnecessary risks, researchers might well prefer the greater convenience offered by a 'disposable species', even if the results are not directly applicable to man. Clinical investigation then, soon became the poor relation of laboratory science and Bernard's fatal doctrine, that animal experiments are entirely conclusive for human patients, duly became enshrined in twentieth-century medicine. Not everyone was happy with the change in emphasis, including Sir Berkeley Moynihan, President of the Royal College of Surgeons:

> 'Much of Galen's work is based upon observation of the

structure of animals; by analogy this knowledge is applied to man. Are not the physiologists today perpetrating the very same error as our great forerunner? . . . Anatomy as a science applied to man was firmly established by Vesalius. Do we need in physiology today a Vesalius to lead us to the true faith? Why, when investigations into the normal processes of physiological activity or into aberrations from the normal can be conducted upon human beings, are so many opportunities neglected? Why are animals selected for the demonstration of certain physiological truths, and why are physiologists content merely to expound truths obtained by observation and experiment, when a visit to the wards would enable the teacher to imprint in indelible characters upon the minds of his pupils these same truths demonstrated upon the human body?'[48]

A year later, in 1928, Sir Berkeley repeated his view that '. . . human physiology has been too much neglected. Research upon animals, for some quite inexplicable reasons, has been put in the foreground.'[49] He blamed the Medical Research Council and the Royal Society '. . . for this misdirection of effort' and argued that neither body was well advised.

But the warnings were ignored and animal experiments became so much the vogue that in new fields of inquiry scientists came to rely on them as the first choice, rather than immediately seeking more reliable approaches or, where applicable, by concentrating on clinical investigation. This often meant that many years passed before alternative means were developed to replace the inevitably unsatisfactory animal-based procedures. Following Pasteur's lead, vaccines were made from living animals or their tissues, often with disastrous results (see also Chapter 6), and it was only when scientists were forced to develop safer alternatives that vaccines prepared from human cells became available.

The Draize test, in which potentially noxious chemicals are sprayed or instilled into the eyes of conscious rabbits, is a good example of how an obsession with animal experiments can delay the introduction of a more humane and reliable approach. The test was introduced in 1944 as a means of assessing eye irritancy but inevitably results proved misleading because the rabbit eye is so different from ours.[50] Scientists themselves repeatedly condemned the test but could only suggest using different species![51] It was only when animal protection groups sharply focussed attention on the Draize test, calling for an urgent programme of research to find an alternative, that attitudes really began to change. As a result there are now several *in vitro* techniques that could replace the Draize test immediately.[52] And, significantly, the technology for all the proposed replacements existed *before* the Draize test itself was introduced. Had animal experiments been prohibited, imaginative scientists would surely have devised similar *in vitro* tests long ago.[53]

Indeed, legal prohibition seems the best incentive of all. In the UK the use of animals to practice microsurgery has been legally prohibited under the Cruelty to Animals Act of 1876 and this has led to the development of the normally discarded human placenta as a viable substitute. The placenta contains tiny vessels that can be sewn together as a means of practice.[54] Blood is pulsed through the system to make it resemble a living being as closely as possible and the efficiency of sewing the vessels together, so essential in microvascular surgery, can be assessed more realistically. In this case necessity was the mother of invention and shows what can be achieved with sufficient imagination, motivation and resources.

Abolishing experiments on animals, then, would not halt medical progress, but force research to concentrate on methods *directly applicable to humans.* As Lawson Tait pointed out over a hundred years ago,

> '... in the interests of true science its employment [vivisection] should be stopped, so that the energy and skill of scientific investigators should be directed into better and safer channels.'[17]

And with the development of tissue culture in the early years of the twentieth century, clinical research would soon have a powerful ally.

1 A. P. Cawadias in *Science, Medicine and History,* E. A. Underwood (Ed.), volume 2 (Oxford University Press, 1953)
2 B. Inglis, *A History of Medicine* (Wiedenfield & Nicholson, 1965)
3 R. Sand, *The Advance to Social Medicine* (Staple Press, 1952)
4 H. Cuny (translated by P. Evans), *Louis Pasteur* (Souvenir Press, 1963)
5 *An Introduction to the study of Experimental Medicine,* C. Bernard, 1865 (translation of the original 1865 text by H. C. Green) (Dover Publications, Inc., 1957)
6 K. Walker, *The Story of Medicine* (Hutchinson, 1954)
7 F. H. Garrison, *History of Medicine* (W. B. Saunders, 1929)

8 D. Guthrie, *A History of Medicine* (Nelson, 1945)
9 C. Wakeley, *Lancet*, 903-906, 9 November, 1957
10 G. R. Davidson, *Medicine Through the Ages* (Methuen, 1968)
11 P. Wingate (Ed.), *Penguin Medical Encyclopaedia* (Penguin, 1976)
12 O. Tomkin, *Galenism* (Cornell University Press, 1973)
13 A. McGlashan, *Lancet*, 1332-1333, 24 December, 1955
14 E. Neil, *William Harvey and the Circulation of the Blood* (Priory Press, 1975). See also reference 11
15 K. D. Keele, *William Harvey* (Nelson, 1965)
16 R. Willis, *The Works of William Harvey M.D.* (Printed for the Sydenham Society 1847, London)
17 Lawson Tait, *Transactions of the Birmingham Philosophical Society*, 20 April, 1882, reproduced in W. Risdon's book *Lawson Tait* (NAVS, 1967)
18 S. R. Riedman, *The Story of Vaccination* (Baily Brothers & Swinfen, 1974)
19 *Daily Express*, 26 October, 1981
20 See Chapter 6
21 See Chapter 1
22 T. Koppanyi and M. Avery, *Clinical Pharmacology & Therapeutics*, 250-270, volume 7, 1966
23 J. T. Litchfield in *Drugs in Our Society*, P. Talalay (Ed.) (Oxford University Press, 1964)
24 R. McGrew, *Encyclopaedia of Medical History* (Macmillan Press, 1985)
25 Z. Cope (Ed.) *Sidelights on the History of Medicine* (Butterworth, 1957)
26 W. A. R. Thomson, *Black's Medical Dictionary* (A. & C. Black, 1981)
27 Reproduced in *Lawson Tait* by W. Risdon (NAVS, 1967)
28 Reproduced in *A Century of Vivisection and Antivivisection*, by E. Westacott (The C. W. Daniel Co. Ltd, 1949)
29 J. Lister, *BMJ*, 377-379, 16 August, 1890
30 In 1869 Coze and Feltz reported *microbes en chainettes* from a case of puerperal fever and Pasteur confirmed this in 1879 by finding the same micro-organism in the blood of another case
31 A. Castiglioni, (1947 edition translated by E. B. Krumbhaer) *A History of Medicine* (Ryerson Press, 1941)
32 According to *Medical Discoveries Who and When* by J. E. Schmidt (Charles C. Thomas, 1959), the Belgian physician Albert Hustin discovered the anticoagulant effects of sodium citrate on blood and advocated its use in transfusions. Physiologists were already aware that citrate had anticoagulant properties but on the basis of animal experiments considered it too dangerous to administer. However, Hustin proved that citrated blood is safe for people by injecting a quantity into a human recipient (*Journal of the History of Medicine*, January, 1954). A year later in 1915, New York surgeon Richard Lewisohn and, independently, an Argentinian L. Agote, published the results of safe transfusions in which patients had been given citrated blood (*Journal of the American Medical Association*, 860, volume 228 1974)
33 Landsteiner reported the discovery of the rhesus factor during experiments with rhesus monkeys in 1940 but a year earlier the same discovery had been made clinically in New York when doctors transfused a young woman after she had given birth (P. Levine and R. E. Stetson, *JAMA*, 126-7, volume 113, 1939)

34 Pasteur's experiments on animals began in the mid 1870s, many years after Lister's introduction of antiseptic surgery in 1866. Pasteur's biographer, Rene Dubois, notes that he began work on anthrax in 1877, chicken cholera in 1879 and rabies in 1880. Dubois writes: 'After 1875 the experimental brewery in the cellar of the laboratory was dismantled to be replaced by a small animal house and hospital, for the study of contagious diseases.' In 1877, Dubois notes, Pasteur published his first studies on animal pathology.

35 *Lancet*, 848-9, 20 March, 1909

36 R. Koch, *BMJ*, 454, 6 September, 1884

37 H. F. Dowling, *Fighting Infection* (Harvard University Press, 1977); see also *Explorers of the Body* by S. Lehrer (Doubleday, 1979) and reference 28

38 B. Inglis, *Fringe Medicine* (Faber & Faber, 1964)

39 'Inequalities in Health', report of a working group, 1980, DHSS

40 *Lancet*, 632, 14 September, 1974

41 *The Advance to Social Medicine* (see reference 3) states: 'The true renaissance of public health dates from 1848 in Great Britain' (see Chapter 1). Pasteur did not even begin his studies on fermentation until 1855 (R. Dubois, *Louis Pasteur*, Victor Gollancz, 1951)

42 Reproduced in reference 2

43 R. D. Ryder, *Victims of Science* (National Antivivisection Society and Centaur Press, 1983)

44 Written by P. Bert in 1878 and included by H. C. Green as an introduction to his translation of Claude Bernard's *Introduction to the Study of Experimental Medicine*, see reference 5

45 *BMJ*, 215, 22 August, 1863

46 C. Singer and E. A. Underwood, *A Short History of Medicine* (Clarendon Press, 1962)

47 G. Hoggan, *Morning Post*, 2 February, 1875, reproduced in reference 43

48 B. Moynihan, *BMJ*, 621, 8 October, 1927

49 B. Moynihan, *Lancet*, 1207-9, 9 June, 1928

50 See Chapter 3

51 For instance, in *Toxicology & Applied Pharmacology*, 701-710, volume 6, 1964, scientists argue that rabbits are unsatisfactory for eye irritancy tests and suggest monkeys instead

52 *Food & Chemical Toxicology*, no. 2, volume 23, 1985 – proceedings of a conference on alternatives to the Draize test, held in Switzerland in April, 1984

53 Several *in vitro* tests for eye irritancy are based on cell culture, a technique that developed since the late nineteenth and early twentieth centuries. Another replacement for the Draize test uses the chorioallantoic membrane of the hen's egg. This has no nerve fibres so cannot feel pain and is rejected by the developing hen's egg, but it can detect inflammation because it has a blood supply and studies have shown it can mimic results from the Draize test (reference 52). The chorioallantoic membrane has been in use for other research since the early years of the twentieth century when it was observed that substances accidentally falling on the membrane caused inflammation. The Draize test itself was not introduced until 1944.

54 *Lord Dowding Fund Bulletin*, no. 23, 1985, see also *The Guardian*, 18 April, 1985

CHAPTER 6

Progress without pain

'The proper study of mankind is man.'

(Alexander Pope)

In Britain during 1879 there were just 270 licensed experiments. Fifty years later there were 403,141 and when in 1970 the figure had reached an enormous 5½ million,[1] it could truly be said that Claude Bernard's charter for twentieth-century medicine had finally come to pass.

By now the obsession with animal-based research had become so great that supposedly reputable scientific journals were publishing the most astonishing experiments, quite apart from the basic scientific objection that results could never be confidently applied to people. In research carried out at the University of Chicago into the importance of sleep, rats were kept awake for up to 33 days. The results were reported in the prestigious journal *Science* and we learn that the animals '. . . suffered severe pathology and death.'[2] Symptoms included debilitated appearance, including yellowed, ungroomed fur, tissue damage, swelling of the paws and severe weakness. Other signs included fluid in the lungs and trachea, collapsed lung, stomach ulcers, severe damage of the scrotum and internal haemorrhage. . . The scientists reached a startling conclusion: 'These results support the view that sleep does serve a vital physiological function.'[2]

Japanese scientists tested common salt for its harmful effects on pregnant mice and found that huge doses (the human equivalent of 5 oz; 140g) given on the tenth or eleventh day of pregnancy actually caused clubfoot in the offspring.[3] *Science* also published research from the University of Oregon in which mice had their forelimbs amputated at birth to see the effect on grooming behaviour.[4] This was part of research into the development of behaviour and it seems the animals still tried to groom themselves with their limb stubs. The journal *Animal Behaviour* reported experiments by Oxford University scientists in which male rats were starved to see the effects on their sexual motivation. 'Under food deprivation', they concluded, 'animals shifted from sexual behaviour to feeding more frequently than when they were non-deprived.'[5] They had made the remarkable discovery that starved animals are more interested in food than sex! Indeed, the field of experimental psychology is full of examples like this, as we shall see in Chapter 7.

The enormous increase in animal experiments over the past hundred years has at least proved one thing: today we have more evidence than ever before that vivisection not only produces little of value compared with clinical research and more modern techniques, it frequently proves misleading.

In his book *Clinical Medical Discoveries*,[6] Dr Beddow Bayly lists many advances achieved entirely through clinical work rather than animal experiments, including anaesthetics, surgical operations to remedy congenital heart defects resulting in 'blue babies', artificial respiration, cardiac catheterization, the use of iodine as an antiseptic and in the treatment of Graves disease, the surgical treatment of unbearable pain, and many diagnostic aids such as the stethoscope, percussion, the electrocardiograph, and the measurement of blood pressure. Surely it is time to concentrate our resources on methods that are not only ethically acceptable but directly related to the *human* condition.

The methods used to advance medical research have become considerably more sophisticated since the nineteenth century and not only include epidemiology and clinical studies but tissue culture and other *in vitro* techniques as well. The causes of human disease can often be discovered by epidemiology in which doctors study entire populations or smaller sub-groups and link disease trends with lifestyle or environmental factors (see also page 57). One of its most striking contributions to health policy in recent years is the discovery that smoking causes lung cancer whilst most of the risk factors linked with heart disease – smoking, high blood pressure, lack of exercise, being overweight and excess cholesterol – were also identified by epidemiology.[7] Studies have shown that atherosclerosis – the build-up of furry deposits in the arteries leading to heart attacks and stroke – is rare amongst vegetarians.[8] And thanks to epidemiology we now know that 80-90 per cent of cancers are preventable.[9]

In the nineteenth century, social reformers like Chadwick used epidemiological methods to influence sanitary reform whilst John Snow's investigation of cholera led to enormous improvements in the quality of water supplies.[7] These are tremendous achievements and, since prevention is always better than cure, the importance of epidemiology cannot be overstated.

The direct study of actual human patients, through clinical observation and research, is another vital part of the jigsaw and, unlike experiments on animals, results are directly applicable to people. In fact, whatever experiments are performed, clinical studies must still be carried out if the disease is to be properly understood: doctors can only hope that animal models do not confuse their clinical findings. And, despite the increasing emphasis on animal experiments over the past century, analysis shows that it is really clinical research that continues to make the important breakthroughs.

Most people believe that Banting and Best, by their dramatic and well-publicized experiments on dogs, produced the cure for diabetes. Setting aside the fact that insulin cannot *cure* the disease, only treat its symptoms, the really important discoveries came through clinical investigation and chemical purification.

The link between diabetes and the pancreas was first demonstrated by Thomas Cawley in 1788 when he examined a patient who had died from the disease.[10] Further autopsies confirmed that diabetes is indeed linked with degeneration of the pancreas but, partly because physiologists, including Claude Bernard, had failed to produce a diabetic state in animals by artificially damaging the pancreas, the idea was not accepted for many years.[11]

Eventually sceptics were convinced when, in 1889, Mering and Minkowski induced diabetes in dogs by surgically removing the entire pancreas.[12] Further autopsies revealed that the islets of Langerhans, a part of the pancreas, were indeed damaged or even completely absent in diabetic patients. If the tissue is damaged, doctors reasoned, patients might improve if given extracts from a healthy pancreas.

There followed a period in which pancreatic extracts were administered to both laboratory animals and diabetic patients, but any beneficial effects were overshadowed by severe toxicity associated with the crude extracts. For instance, in 1908 Zuelzer administered pancreatic extracts to patients and their diabetic symptoms actually improved[13], but the tests had to stop because of undue toxicity. Even after their experiments with dogs 13 years later, Banting and Best's first human trial proved disappointing and had to be stopped with Banting himself admitting that 'results were not as encouraging as those obtained by Zuelzer in 1908.'[14] Only when the biochemist Dr J. B. Collip succeeded in purifying the extracts did a more effective and less toxic preparation become available. Ultimately this was the crucial step in providing a relatively safe treatment for insulin-dependent diabetes.

Today we know, by comparing diabetic patients with non-diabetics, that the more common maturity onset diabetes can often be prevented and in most cases treated with diet alone. Recent research has also shown that most cases of insulin-dependent diabetes are preventable provided that the key environmental factors can be determined.[15] But in the absence of preventive action, and because the disease has an hereditary component, the effect of the discovery of insulin was to cause a steady rise in the number of diabetics, as patients survived long enough to have children of their own.[16]

Another case is the treatment of amblyopia, a defect of vision called 'lazy eye' that sometimes approaches blindness. Inevitably the human condition has been used to justify any number of sight deprivation experiments on animals:

> 'Work on the development of the mammalian visual system may lead to exploration of human visual disorders and to the *identification of sensitive periods during which the visual system is capable of modification*. . . The programme grant held by Professor H. B. Barlow and Dr C. B. Blakemore . . . has been renewed. Basic research in this field should facilitate the investigation of more applied problems such as the treatment of amblyopia. . .'

(emphasis added)
(Medical Research Council Report, 1977)

Yet, by the time the first sight deprivation experiments were reported by Wiesel and Hubel in 1963, several clinical studies had *already shown* that there is a critical period in the development of the human visual system.[17] According to Dr Drewett of Durham University's Department of Psychology,

'... the fact that there is a critical period in the development of vision, and its clinical implication, that visual defects should be detected and corrected as early as possible, did not derive from this work on animals. It was already known.'[18]

Clinical investigation has also proved invaluable in brain research, both in neurology – the study of diseases of the nervous system – and in the localization of cerebral functions. By clinical observation and post-mortem examination of patients with brain injuries, doctors have been able to relate changes in behaviour to specific regions of the brain. For instance, as long ago as the early 1800s, doctors realized that in people who have lost the power of speech during life, the brain shows signs of disease in the left frontal lobe after death. In 1861, Broca made the first definite discovery of cerebral localization by proving that the faculty of speech is governed by a centre in the region of the inferior frontal gyrus, named Broca's convolution.[19]

Later on, between 1863 and 1870, the great clinical researcher Hughlings Jackson of the London Hospital observed that certain forms of epilepsy are caused by disease affecting the part of the brain bordering on the central sulcus. Only later were Jackson's observations confirmed, as far as they could be, in animal experiments.[20] And it was through mild electrical stimulation of the *human* brain that doctors expected to produce a reliable map of the sensory functions.[21]

Before the use of animals reached epidemic proportions most neurological work was based on clinical studies, doctors recognizing, as pointed out by Sir Russell Brain in his review of the neurological history of the London Hospital, that only the study of disease in man can alleviate the problems of human illness.[20]

In 1960 Ehringer and Hornykiewicz discovered that patients with Parkinson's disease had depleted levels of a chemical called dopamine in those regions of the brain most affected by the illness. This led to the successful use of L-dopa, a drug converted into dopamine in the brain, thereby making up the deficiency.[23]

Today sophisticated new techniques are enabling research with volunteers and patients to be carried out more safely, giving new insights into the disease process. One example is a remarkable imaging technique called positron emission tomography. Using a minute amount of radioactive chemical to mark the brain's active areas and a brain scanner to detect the chemical, pictures are produced that show the brain in action – in health and disease. Techniques like this are being used to study stroke, coronary artery disease, epilepsy, Parkinson's disease and others.[24] When artificially induced in animals, the disease can take quite a different form.

Often it is chance observation rather than planned research that provides an alert physician with the vital clue. Crawford Long's discovery of the anaesthetic properties of ether is a good example[25] of this and another famous case is the early study of stomach physiology by the United States' army surgeon William Beaumont.[26]

In the 1820s Alexis St Martin, a young Canadian trapper, received a gunshot wound in the abdomen and thus came under Beaumont's care. Although the wounded man eventually recovered, he was left with a gastric fistula and through this artificial opening Beaumont was able to observe the walls of his patient's stomach and obtain pure gastric juice. For two years the patient and surgeon became partners in a valuable piece of physiological research with Beaumont making almost daily observations and experiments. Chemical analysis showed that the gastric juice contained free hydrochloric acid and was only secreted when food entered the stomach. Beaumont demonstrated the action of gastric juice on different foods both within the body and *in vitro*, and carefully noted the changes in stomach physiology as a result of fear and anger, feverish symptoms and excessive alcohol intake. Ultimately Dr Beaumont worked out over 50 positive conclusions and according to Sir Arthur Hurst, Senior Physician to Guy's Hospital, his research '. . . laid the foundation of our knowledge of gastric digestion.'[27]

Clinical research can also use tissue from patients, for instance in the study of arthritis, cancer and diseases of the nervous system. Tissue culture techniques enable cells and tissues to be kept alive outside the body and have the advantage that *human* tissue can be used to avoid the problem of species variation. Furthermore results can be directly matched with clinical and epidemiological findings, providing an overall picture of the disease process. Tissue is obtained from healthy volunteers, during therapeutic surgery or from autopsy specimens and can often be kept in cold storage until required. Although for convenience the method is usually referred to as *tissue culture*, there are, in fact, two main categories: *organ culture*, in which small pieces of tissue are maintained in a nutrient medium so as to maintain the function of the organ from which they are removed, and *cell culture*, in which dispersed cells are cultured in a medium allowing continuous growth. Organ culture is obviously more closely related to the situation in the body and a good example of its use is in cancer research. For instance, organ cultures derived from individual human pituitary tumours continue to synthesize and secrete hormones just as they do in the living person and they also respond in the same way when stimulated by chemicals found in the body.[28]

In research carried out by scientists at America's Food and Drug Administration, an organ culture of human muscle tissue, normally discarded after routine surgery, has formed a test bed to study the growth and spread of human cancer cells.[29] The system successfully mimics the situation in the body where cancer cells proliferate, spread and invade the surrounding tissues and, according to the researchers, '... offers the possibility of studying not only the biology of cancer cell growth and invasion into normal human tissue, but also provides a method for evaluating the effects of a variety of potentially important antitumour agents such as interferon, lymphocyte sub-populations, and antibodies.'[29] Anticancer drugs could therefore be tested for their effects on tumour cells *and* the host tissue:

> 'The human muscle system provides for the first time an *in vitro* experimental model using easily accessible adult human tissue to study cancer and its treatment.'[29]

Because of the many different forms of cancer, other experts have recognized that research on tumour cell properties,
> 'must concentrate more on the diverse nature of human tumour cells recently removed from patients and less on animal model systems.'[30]

Yet the leading medical journal *Cancer Research* reveals that most of its published articles in basic cancer research involve animals or their tissues.[31]

Another example of organ culture is in cataract research where tissue is obtained from patients undergoing routine surgical treatment for the disease. Recent findings suggest that the lens from the human eye behaves quite differently to those from other species.[32] Cell cultures too can give valuable insights: using cultures of human cells derived from hysterectomy specimens, scientists at Hammersmith Hospital are investigating infections of the female genital tract. According to the researchers:

> 'The cultures provide an alternative approach to the study of invasion of tissues of the female genital tract by pathogenic micro-organisms. They are more relevant to humans than animal experiments, are more economical and produce results quickly. Direct observation of the pathogenic processes involved is easy using scanning electron microscopy.'[33]

Scientists at the Middlesex Hospital and Medical School have recently isolated individual heart cells from human heart muscle. The cells are expected to prove useful not only for research into heart disease but into the preservation of heart tissue for cardiac surgery. And there is the added advantage that results are directly applicable to patients since, as the researchers explain, '... it is difficult and often misleading to extrapolate experimental results in animal tissues to man.'[34]

Drug research

Many of our most important drugs were first discovered through clinical studies with patients or entirely by chance, as in the case of penicillin. In the past the use of digitalis for heart disease, quinine to fight malaria, ipecac for the treatment of amoebic dysentery, the use of narcotic agents such as morphine to relieve pain, caffeine as a central nervous system stimulant, veratrum alkaloids for lowering blood pressure, iron for the treatment of anaemia, ergot alkaloids for contracting the uterus and relieving migraine, the use of belladona alkaloids (which contain the important drug atropine) and salicylates, and the development of the early inhalation anaesthetics, were all discovered through work with human patients.[35]

Many of these early drugs are still important remedies that have stood the test of time: quinine is still used to treat malaria whilst digitalis and related compounds remain the mainstay of modern treatment of heart failure; morphine is vital in coping with severe pain and the anaesthetics ether and nitrous oxide, discovered during the nineteenth century, are listed as essential drugs by the World Health Organisation.[37]

Ether is considered to have a wide margin of safety,[38] unlike chloroform which was later replaced by safer, chemically related, alternatives. The anaesthetic effects of Trilene, a drug closely related to chloroform but considerably less toxic,[39] were also discovered by chance through its use as a cleansing agent in industry: workmen learnt the effect of inhaling the vapour when they leaned over degreasing vats containing the liquid. Trilene has since proved useful mainly during short surgical procedures where light anaesthesia with good analgesia is required, as in obstetrics.[39]

Stressing the need for astute clinical observation, John Litchfield, Director of Research at Lederle Laboratories, describes many other drugs whose beneficial effects were first discovered in human beings and only later confirmed in laboratory animals.[40] His more modern examples include the diuretic effect of sulfanilimide, the tranquillizing action of chlorpromazine, the analgesic effect of methotrimeprazine, the ability of thiazide diuretics to lower blood pressure, the use of cortisone to relieve arthritis and the discovery that fluoride can protect against dental caries. Even in cancer research with its traditional emphasis on laboratory animals, most of the important drugs originated from clues derived during clinical investigation.[41]

In a Congressional Testimony during 1981 Dr Irwin Bross, Director of Biostatistics at Roswell Park Memorial Institute for Cancer Research in Buffalo, New York, wrote:

'The uselessness of most of the animal model studies is less well known. For example, the discovery of chemotherapeutic agents for the treatment of human cancer is widely-heralded as a triumph due to use of animal model systems. However, here again, these exaggerated claims are coming from or are endorsed by the same people who get the federal dollars for animal research. There is little, if any, factual evidence that would support these claims. Indeed, while conflicting animal results have often delayed and hampered advances in the war on cancer, they have never produced a single substantial advance either in the prevention or treatment of human cancer. For instance, *practically all of the chemotherapeutic agents which are of value in the treatment of human cancer were found in a clinical context rather than in animal studies.*'[42] (emphasis added)

Doctors now know that one of the most important methods of discovering a drug's beneficial effects is, paradoxically, through an analysis of its side-effects. These need not be harmful, just additional to the main effect. For instance, clinical studies may reveal that a drug reduces blood pressure in addition to its main function. These 'unwanted effects' show that the drug might also be used to treat high blood pressure. Indeed, so productive has the method been that Professor A. D. Dayan, formerly a leading toxicologist at Britain's Wellcome Research Laboratories, has referred to the '. . . humiliatingly large number of medicines discovered only by serendipitous observation in man (ranging from diuretics to antidepressants), or by astute analysis of deliberate or accidental [human] poisoning.'[43] For instance, three out of the four major classes of drugs now used to treat high blood pressure were not known to have this effect until they were given to patients for other conditions. According to Dr Franz Gross, one of the world's leading researchers in the field of high blood pressure,

> 'It has to be admitted that the antihypertensive effect of some drugs such as the diuretics, clonidine, or the Beta-adrenegic blockers [beta-blockers] were first observed in man, and only later were studied in animal experiments with respect to their blood pressure lowering activity. It is however, doubtful if their therapeutic effects in hypertension had been detected in the laboratory on the basis of a larger screening programme in rats with different types of experimental or spontaneous hypertension.'[44]

Major disease conditions treated by drugs not originally introduced for that purpose[45]

Drug	Condition treated
Propranolol	High blood pressure
Sulphinpyrazone	Heart disease
Diazepam	Status epilepticus
Phenobarbitone	Epilepsy
Chlorpromazine	Schizophrenia
Oestrogens/progestogens	Contraception
Imipramine	Depression
Probenecid	Gout
Lignocaine	Arrhythmias

Thus the blood pressure lowering effects of drugs that constitute the mainstay of treatment for hypertension – the diuretics and the beta-blockers – were discovered through their effect in patients. More examples are listed in the table opposite.

Another case is the use of drugs to suppress the body's natural defence mechanism during organ transplants, an essential element if the operation is to prove successful. Many of the drugs commonly used for the purpose such as steroids, had long been used to treat cancer when it was discovered through clinical observation that they also adversely affected the immune system, leaving the body open to infection.[46] Inevitably countless animal experiments have since taken place but the essential discovery that led to the 'successful' development of transplant surgery can be directly traced to work with human patients.

Even if animal experiments were stopped tomorrow and there were no other alternatives in view, observations of this kind, based on the astute analysis of side-effects, would still lead to important advances as they have in the past.

Traditionally the pharmaceutical industry has identified new drugs by testing a vast range of chemical substances on artificially-induced animal models of the disease. Ciba Geigy estimate that out of 3,000 chemical compounds tested, only 20 show enough therapeutic activity and low enough toxicity on the basis of animal experiments, to be tested on human volunteers. Of these 20 only one ever becomes a prescription drug![47]

As we have seen, most new medicines add little to those already available but, if such testing programmes are to continue, there are alternative methods that could reduce or replace the use of living animals. Professors Farnsworth and Pezzuto of the University of Illinois' College of Pharmacy argue that sufficient *in vitro* techniques now exist so that almost any useful drug effect can be predicted without using animals.[48] Examples include the use of enzymes to predict blood pressure lowering effects whilst isolated blood platelets can be used to find drugs for the treatment of cardiovascular problems.

The marked lack of success in producing effective new anticancer drugs has prompted several doctors to suggest the use of *human* cancer cells as an alternative to animal experiments.[49] The cells could come from several major types of the disease (eg, cancer of the lung, colon, melanoma, ovary) so that useful drugs are less likely to be missed. After testing *in vitro*, promising drugs would bypass any animal cancer tests as they might only cause confusion, and go directly to the clinic. And it is reassuring to know that such tests are capable of mimicking the patient's response: with relatively few exceptions, those agents known to be effective in treating human cancer are also active in cell culture tests.[50]

Recent reports suggest that the US National Cancer Institute has finally decided to change the way it searches for new anticancer drugs. At present all compounds are tested on animals with leukemia but in the new system, substances will be tested *in vitro* against more than 100 different strains of human cancer. Officials believe the new method will be far more sensitive than the old one, identifying promising drugs that would have been dismissed as useless in the past.[51] Unfortunately, new drugs passing the first stage will then go on to animal tests where misleading results can again spell disaster.

Potential new antibiotics, often derived from the soil as in the case of streptomycin, can be identified *in vitro* by their ability to kill test organisms such as *Eschericha coli*. In the past promising candidates would then go on to tests with deliberately infected animals, although the final test was always on man. But now researchers at the University of Leeds have developed a promising alternative.[52] Micro-organisms regularly gain access to the body's tissues but rarely cause infection due to the effective clearing action of the host defence mechanisms. Most bacterial strains are killed when incubated with serum from a normal person and this effect is due to a complex series of proteins called 'complement'. Special white blood cells (phagocytes) are a second major host defence against invading bacteria: these cells localize at the site of infection where they capture, ingest and destroy any foreign bodies. The new *in vitro* tests developed at Leeds, by using human serum (containing complement) and white blood cells (phagocytes) now have a direct bearing on the behaviour of drugs and bacteria within the *human* host.

Since the discovery of penicillin, several hundred thousand synthetic and natural products have been screened for antibiotic activity and, of those subjected to clinical trials, only one in 50 would be considered suitable for general use.[53] However, drugs submitted for clinical trials must presumably have cured infected animals but a 2 per cent success rate is a terrible indictment of the preliminary animal tests and argues strongly for a more reliable approach.

Like antibiotics, antiviral agents can also be screened *in vitro*, potential new drugs being tested for their effects on viruses like influenza, herpes and AIDS. In the search for drugs to treat influenza, scientists at ICI have used both cell and organ cultures to dramatically cut the number of animals originally used.[54]

The influenza virus attacks the lining of the trachea, which results in the characteristic respiratory symptoms. By using cultures of fragments of trachea, the process of virus infection and tissue damage can be followed in the test tube, and drugs tested for their effectiveness in preventing virus growth. Although the organ cultures come from the trachea of animals, there is no good reason why human tissue could not be used instead.

In the early 1960s, the ICI laboratory used over 12,000 mice a year to test less than 1,000 compounds for antiviral activity but, by 1977, as their dependence on tissue culture systems grew, over 22,000 substances were being tested and the number of mice used had fallen to about 2,000.

Alternatives have been suggested for testing drugs to treat herpes virus infection of the cornea. Not surprisingly, living animals, particularly rabbits, have been used to investigate the disease, being inoculated with virus directly into the eye. Experiments carried out by the Wellcome Research Laboratories reveal the true horror of these tests.[55] After three days, when ulceration at the site of infection was well developed, one eye was treated with the test drug and the other was left untreated as a control. 'In the untreated eye ulceration progressed . . .'[55] until the injuries all flowed together. 'At this stage the eye was grossly inflamed with . . . purulent discharge, opacity of cornea and extensive ulceration.' *The experiments continued for seven days and there is no mention of any pain relief.* Other scientists, referring to the 'severe occular lesions' seen in such tests, have now suggested an alternative approach using isolated eyes taken from animals killed for food.[56] Whilst the test still depends on animal tissue, it does avoid the pain and suffering experienced by living animals and could be improved still further by relying on human tissue from eye banks.

Rapid screening of drugs against the AIDS virus can also be carried out in the test tube.[57] Clinical observation has shown how the virus works so enabling a simple *in vitro* test to be developed. The AIDS virus disturbs the growth of the T4 lymphocyte, a cell crucial in regulating the body's defence mechanism. When the T4 lymphocyte is damaged, patients become open to life-threatening infections that their bodies would otherwise resist. The beneficial effects of new drugs are therefore easily assessed by adding them to cultures of the T-cells to see if they protect against damage caused by the virus. Already drugs active *in vitro* are being submitted for clinical trial.

In addition to *in vitro* tests there is growing interest in predicting the useful biological effects of a substance on the basis of its chemical and electronic structure. One of these techniques is quantum pharmacology, where the behaviour of drugs can be explained by mathematical calculations involving the energy levels in chemical substances.[58] Such approaches, which increasingly rely on computer graphics, can identify promising drugs without any preliminary animal experiments. Designing drugs by computer is a far cry from the hit and miss, mass-testing programmes traditionally employed by industry.

Safety testing

Comparisons between human and animal test data reveal that most drug side-effects cannot be predicted by animal tests at all.[59] Animal experiments simply do not have the potential to predict most of the adverse reactions arising in patients and even when they do, results are often misleading. Since the true dangers and benefits of medicines only really emerge after wide use by human beings, it seems more sensible to concentrate on careful monitoring of patients during clinical trials and after a drug is marketed than to bring suffering and death to animals in tests of little relevance.

> 'We know from drug toxicity studies that animal tests are very imperfect indicators of human toxicity: only clinical experience and careful control of the introduction of new drugs can tell us about their real dangers.'
>
> (*Lancet*, 1972[60])

But at present there is little effective monitoring of drugs in human patients once a product has been marketed. The main UK post-marketing surveillance scheme, where GPs voluntarily make reports on special yellow cards, as noted previously, only reveals 1-10 per cent of drug side-effects[61] and even very serious hazards are grossly under-reported: only 11 per cent of fatal reactions associated with anti-inflammatory drugs phenylbutazone and oxyphenbutazone, were actually reported. Even worse only about a dozen of the 3,500 deaths linked with isoprenaline aerosol inhalers during the 1960s were reported by doctors.[62] By neglecting proper monitoring systems doctors are, in effect, relying on the initial clinical trials together with the results from animal experiments to protect patients once a drug is marketed. And whilst clinical trials reject the vast majority (95 per cent) of drugs passed safe and effective on the basis of animal tests,[47] they do involve relatively small numbers of people, so many unforeseen hazards only appear once a drug is marketed and widely used. In order to prevent the continual spate of drug disasters, reliance should, therefore, be switched from misleading animal tests to an effective system of postmarketing surveillance, in which patients are properly monitored whilst new drugs are slowly and carefully introduced onto the market. Of course, reliable nonanimal techniques could be used to weed out the worst offenders prior to tests with volunteers.

An effective postmarketing surveillance scheme would have quickly detected the eye damage caused by the heart drug Eraldin and prevented a major disaster. The drug was prescribed for over four years before doctors realized its terrible effects. Then, within weeks of the first published report, 200 more cases came flooding in as physicians were alerted to the danger.[62] Eventually ICI compensated more than 1,000 victims, yet the original animal tests had given no hint of the tragedy to come.

Robert Brent from the Department of Paediatrics at Jefferson Medical College, believes that efficient clinical surveillance schemes would have uncovered the link between thalidomide and limb malformations after only a handful of cases, so preventing a major catastrophe.[64] Whilst this would still have been too many, potent drugs will always have their darker side, stressing the urgent need to avoid drugs during pregnancy except for the most pressing indications.

In 1977 European scientists met to discuss these questions, arguing that '. . . present methods of testing for safety consume time, money, and manpower, without a corresponding increase in safety.'[65] They concluded:

'We must recognize that existing methods are unsatisfactory. We recommend more rational and less extensive laboratory

studies, without unnecessary multiplication of detailed clinical
trials before registration. Instead we recommend much closer
and more extensive surveillance of medicines after they are
available for general prescription. Only by the careful study of
medicines in everyday use can the greatest benefits be
obtained from their administration, the untoward rare
potential disaster be recognized at the earliest possible
moment, and the ill effects be minimized.'[65]

Before new products are submitted for clinical trial, some
preliminary testing is essential but humane alternatives are
possible. A New York company, Health Designs Incorporated,
has developed computer programmes to predict the results from
a variety of toxicity tests without using animals. Existing data is
used to develop mathematical equations whereby toxicity can be
predicted on the basis of chemical structure. For 80 per cent of
the test chemicals the model successfully predicted the LD50
within a factor of ± 6, which is well within the variation (8-14x)
that can occur between individual laboratories when animals are
used to measure the LD50. The researchers concluded that,

> '. . . the level of accuracy of the estimates produced by the
> model is similar to that obtainable from animal bioassays.'[66]

Furthermore the results do *not* imply that LD50s for 20 per cent of
the chemicals were predicted inaccurately. In several cases, where
the actual and predicted LD50 values differed considerably, it was
found on careful re-examination that the original animal test
results were incorrectly reported in the scientific literature. If a
numerical index of toxicity is still to be required, say for
classification purposes, why not rely on the computer model
instead of animal experiments? And why not develop the same
programme to predict *human* lethal doses? In any case the
enormous differences that often occur between lethal doses in
animals and humans means that LD50s should never be used as a
guide to dose levels for volunteers in clinical trials. For the safety of
those taking part, such trials must always commence with minute
amounts of the drug, whatever the preliminary tests say.

In another example, the technique has been used to predict
skin irritancy with excellent results.[67] Traditionally, tests with
rabbits are used to classify chemicals as severe, moderate, mild
or nonirritant. Health Designs have shown that for 91.5 per cent
of the chemicals tested the programme successfully distinguished severe irritants from the remainder whilst for 93 per cent
of the substances, the technique correctly distinguished

nonirritants from mild/moderate skin irritants. The differences between human and animal skin have led to the suggestion that substances of low or medium skin irritancy could be tested on human volunteers instead of animals.[68] Preliminary tests, perhaps using cultures of skin tissue or the theoretical methods devised by Health Designs could be used to weed out severely irritant chemicals.

Health Designs can also predict whether chemicals are likely to cause cancer or birth defects and other scientists at the Johns Hopkins University in Baltimore are designing computer programmes to predict general toxic effects.[69] Theoretical techniques can be supplemented with *in vitro* methods such as tissue culture to assess various aspects of toxicity, including cancer, birth defects, skin and eye irritation, and the effects of pathogenic mineral dusts like asbestos. For instance, eye irritants can be identified using cell or organ cultures as an alternative to live rabbits in the Draize test.[70] In another example, a simple cell culture correctly estimated the human lethal dose for 75-80 per cent of substances tested, indicating that most chemicals cause injury by interfering with the basic working of the cell.[71] On the other hand some chemicals are toxic to specific organs and a more sophisticated *in vitro* test would include tissue from those commonly affected such as the liver, kidney, skin and nervous system. For example, several drugs like chloramphenicol are known from clinical experience to damage the bone marrow, producing deadly blood diseases. Now, scientists at the Lister Hospital in Hertfordshire have suggested the use of human bone marrow cultures to detect dangerous drugs before they are submitted for clinical trial.[72] They argue that any test-tube method using human tissue gives a degree of reassurance *not* provided by animal experiments or by procedures using animal tissues. In another case, cultures of human kidney cells have been proposed to evaluate the damaging effects of aminoglycoside antibiotics. These drugs are known to cause kidney problems in 2-30 per cent of patients and are normally tested on animals to assess kidney damage. However, tests with human kidney cells correctly ranked known aminoglycoside antibiotics according to their effects in patients.[73]

Nonanimal tests to discover whether chemicals cause cancer have developed rapidly because traditional animal cancer tests are very costly, using around 600 animals per test and taking up to three years to perform. They cannot possibly cope with the 40,000 (largely untested) chemical compounds currently in use in our environment.

The best-known alternative is the Ames' test, which exposes salmonella bacteria to substances suspected of causing cancer.[74] The technique, which is highly sensitive and 20 times cheaper than animal tests, only takes a few days to perform and is based on the idea that cancer producing chemicals will identify themselves by mutating genes in the bacteria. In laboratory trials the method gave a 90 per cent success rate in predicting the carcinogenicity of more than 300 chemicals and is now used in over 2,000 laboratories. Unlike laboratory animals, the method readily detects the hazardous nature of cigarette smoke. Nevertheless, the test cannot detect all dangerous substances and should therefore be used in conjunction with other, similar *in vitro* techniques that use yeasts or cell cultures. Finally, there is the danger to society that continued reliance on expensive and time-consuming animal cancer tests could result in a backlog of chemicals that would never be tested.

Commonly animal researchers complain that *in vitro* tests cannot mimic effects on the whole body, as if animal experiments are always correct in predicting human responses! In fact, both methods have great limitations but tissue culture does have one great advantage in that *human* material can be used and the problem of species variation avoided. It is pointless having the complexity of the whole animal if results continually prove misleading. And, significantly, thalidomide's notorious toxic effect can indeed be seen in human tissue culture[22] but not in rats and mice. Furthermore, ingenious scientists have found a way to make *in vitro* systems more closely resemble the living person. Sometimes chemicals only become hazardous when metabolized, usually in the liver, so researchers include liver cells in their *in vitro* tests to mimic the body's metabolic processes. The Ames' test routinely includes cells from the rat's liver although more relevant results can be expected with human liver cells.

But, whatever preliminary experiments are carried out, the first really valid test of a new drug comes when it is given to healthy volunteers and patients during carefully controlled clinical trials. *In vitro* tests based on human tissue at least offer the prospect of predictions that directly apply to the patient and, as scientists point out, give a degree of reassurance not provided by animal experiments.

Training

Most scientists using animals regard them as just another research tool, part of the standard laboratory equipment. The conditioning begins at an early age when children are required to witness or carry out dissections at school. The Institute of Biology has said that there is no need to perform dissections before the age of 16, whilst some universities, including Oxford and Cambridge, accept students for biology courses without them studying biology at A-level, let alone carrying out dissection.[75] So dissection in schools cannot be 'essential'. Models, diagrams and films could be used instead. After all, human anatomy and biology is taught at schools without dissecting a *human* corpse!

Scientists at the University of Bradford are using a Biovideograph instrument, developed by Bioscience, as an alternative to animals for teaching pharmacology and physiology. The Biovideograph is a particularly sophisticated form of video recording where students not only watch the experiments on film but actually record the experimental results themselves. This gives them a degree of participation not provided by ordinary video recordings. The instrument itself is connected to the usual video machine and TV set, and additionally gives a playback of experimental data recorded by a pen on a chart recorder. Each group of students therefore finishes the 'experiment' with their own permanent record of physiological data such as blood pressure or heart rate. One example showed how the method could replace the repeated use of cats for the demonstration of blood pressure effects of certain drugs.[76] Professor Leach, former chairman of the School of Studies in Pharmacology at the University of Bradford, has taught a complete experimental pharmacology programme by this method. In fact studies by Henman and Leach show that the method is not only quicker and cheaper than conventional class experiments with actual animal tissues, but provides a better standard of learning performance.[63]

But reliance on traditional practices means that such techniques are underutilized. Fortunately pressure is coming from the students themselves. In 1986 the then NAVS Organiser Lorraine Walker and student Judith Anstee initiated a campaign towards Violence-Free Science, aimed at giving students the right to science education without animal

190 *The Cruel Deception*

exploitation. The campaign is based on a special charter that protects students from academic penalty should they object to animal experiments during the course of their studies. Already student unions from over 20 universities, polytechnics and colleges have backed the charter.

Sometimes the cost of animals overcomes entrenched attitudes. An internal application for equipment grant by the department of Physiology at the University of Leeds states:

'. . . we are making increasing use of the students as subjects in neurophysiological experiments because cats are so expensive to buy and in increasingly short supply because of antivivisectionist propaganda. However, some new equipment is required because that used in animal experiments does not meet the safety standards relating to apparatus for patient use (principally the avoidance of electric shock hazards).'[77]

The use of computers to reduce or replace living animals in medical schools has numerous advantages according to one of the technique's pioneers, Dr James Walker of the University of Texas:

'. . . computer simulations offer a wide range of advantages over live animal experiments in the physiology and pharmacology laboratory. These include: savings in animal procurement and housing costs; nearly unlimited availability to meet student schedules; the opportunity to correct errors and repeat parts of the experiment performed incorrectly or misinterpreted; speed of operation and efficient use of students' time and consistency with knowledge learned elsewhere.'[78]

The 'Mac' family of interactive digital computer programmes are simulation models for use in clinical, physiological and pharmacological teaching and research.[79] The student can monitor important physiological variables such as heart-rate, blood pressure, arterial oxygen and carbon dioxide pressures, ventilation, acid-base status, urine output, plasma potassium, haemoglobin and plasma concentration of drugs, and by altering one or more of these variables he can study the effects on the physiological system being simulated over a period of time. The progress of the simulated patient is displayed graphically at the computer terminal and symptoms are printed at regular intervals. These programmes originated from MacMaster University in 1970 and have been under continuous development ever since – most recently at St Bartholomew's Hospital Medical College in London, where they can be obtained for educational and research purposes. The usual objection to computer simulations is that they do not provide experience of working with living beings, and, ultimately, medical students need to learn about disease at the bedside, as Hippocrates taught.

Because of the UK's legal prohibition on the use of animals to practice surgical skills, surgeons have learned their craft by practising on human bodies in the mortuary, then by observing senior surgeons at work and, finally, by operating themselves under the close supervision of experienced colleagues. We have already seen how surgeons at Frenchay Hospital in Bristol have developed the normally discarded human placenta as an alternative to animals for training in microsurgery. According to Paul Townsend, the Consultant plastic surgeon who pioneered the work:

> 'Due to the problems of reproducing pulsatile blood flow, training in microvascular anastomosis [combination of vessels] has generally been restricted to animal work. However the use of the human placenta eliminates ethical concern as well as providing a unique source of human material without danger to the patient. Appropriate sized vessels which are easily seen on the surface of the placenta can be selected, dissected, divided and re-anastomosed. By providing a dynamic artificial circulation with pulsatile flow, using saline or blood, a far more realistic model is available.'[80]

Ironically, the legal prohibition on the use of animals to practise surgical skills, which led directly to the placental model, has now been lifted but it will hardly encourage the technique's widespread use. The Home Office argues that the placenta does not contain nerve fibres and so cannot be used to practice nerve sutures. This sounds a feeble excuse because the placental model does provide the surgical skill necessary to work through a microscope at a minute level, even though there are no nerve fibres. In any case, as Mr Townsend has explained, compared with joining nerve fibres, it is the proper reconnection of blood vessels that is crucially important in microsurgery.[80]

Vaccines and other biological products

Apart from his announcement of the germ theory, Pasteur's other great claim to fame is the development of the first vaccine against rabies. As his biographer René Dubois pointed out, it seems strange he should have chosen rabies, an extremely rare disease amongst humans, even when prevalent in surrounding animals. According to the World Health Organisation there were only 23 cases of human rabies in the whole of Europe,[81] including Turkey, between 1977 and 1982 whilst just 16 cases have been reported in North America since 1966.[82] In Europe at least, deaths from the disease remain much fewer than from lightning strikes[81] and when a patient died in Oklahoma during 1981 it was the first case of human rabies for two years in the United States.[82]

Pasteur's vaccine, made from deliberately infected brain tissue of living animals, has always been controversial and not only for its dangerous side-effects. Clinical reports indicate it simply does not work when injected after a rabid bite.[83] Today we know that few people bitten by a rabid animal actually develop the disease, which means that those who 'recovered' after a course of inoculations may not have been infected in the first place. In fact, according to a British government Memorandum on Rabies, '. . . man is not highly susceptible to the disease.'[84] The chances of contracting the disease depend on the severity of the bite and where the patient is bitten but generally rabies develops in a small fraction of untreated cases after bites from rabid dogs.[85]

By 1973, when the World Health Organisation announced its sixth report on the control of rabies, vaccines made from infected brain tissue were still the most extensively used. The WHO recommended that Fermi-type vaccines should no longer be used because *they still contain residual live virus*! And whilst live virus had been eliminated from the Semple type vaccines, these still proved hazardous because humans are allergic to brain tissue from other animals: vaccination illness varying from mild local reactions to severe neurological damage including paralysis and death. So dangerous were they that Pennsylvania rabies researchers Klaus Hummeler and Hilary Koprowski pointed out as recently as 1969 that vaccines made from brain tissue are '. . . still the worst biological products ever injected into the human body.'[86] A far safer vaccine made from cultures of *human* cells avoids these problems and succeeds in developing antibodies to the disease in human patients.[87]

The most valuable procedure after being bitten by a suspected rabid animal, stresses the WHO, is the local treatment of wounds, which should include thorough washing with soap and water together with the use of antirabies serum containing antibodies to the disease. Antirabies serum derived from horses, mules and donkeys often causes serum sickness and the WHO emphasises the advantages of human antirabies serum.

But if Pasteur's vaccine turned out to be a failure when injected after a rabid bite, his laboratory methods were to prove far more influential. Following his example, scientists used *living animals* to develop the first vaccine against poliomyelitis, and once again the results were disastrous. They had first demonstrated that the spinal cords of baboons, chimpanzees and rhesus monkeys could be infected with the virus. Then, after killing the animals, the infected spinal cords were ground up; attempts were made to render the virus harmless and the mixture was finally injected back into monkeys to see if the infection had lost its virulence. Success in laboratory animals led to the first attempts at human vaccination by Kolmer and Brodie in 1934/35 but the trials ended in disaster and there were several deaths.[88]

By far the most important advance in the development of polio vaccine came in 1949 when Enders, Weller and Robbins showed that all three main types of polio virus could be grown in human tissue culture. Leonard Hayflick of Stanford University's School of Medicine and a leading figure in the development of safer vaccines, writes:

'The immediate result of this observation was the development of effective vaccines against poliomyelitis, but the impact on virology resulting from the realization that cell cultures were largely to replace the laboratory animal and embryonated eggs was much more far reaching.'[89]

The virus was soon shown to grow in many human tissues including muscle, intestine, foreskin, spleen and heart but monkeys were obviously a far more convenient source and, in 1956, Salk developed a killed vaccine using kidney tissue derived from monkeys. And later on, when Sabin produced a live attenuated vaccine, he too decided on monkey kidney tissue. It was to prove an unfortunate choice, for monkeys are known to harbour more than 60 viruses, some of which are dangerous to humans.[90] Several hundred thousand people at least have been inoculated with live SV40 virus found in polio vaccine produced in monkey kidney cells. The virus has the potential to cause cancer.[91] Furthermore, the cancer-causing viruses that contaminate tissue from other species, only become dangerous when they cross the species barrier.[89] Marburg agent, a hitherto unknown virus, and Herpes B virus have also been found, causing several fatalities.[89] Indeed the effect on those handling the monkeys and their tissues was far more immediate and by 1972 23 people had died. Furthermore contamination with harmful viruses meant that 60-80 per cent of cells had to be rejected, which resulted in even more monkeys being killed.[92] And there is still the ever-present possibility of unknown and undetectable viruses finding their way into the vaccine, with potentially catastrophic results. As P. B. Stone of Pfizer's Therapeutics Research Division warned in 1970:

'There is . . . always the prospect of a new or unknown virus which would not be detected by the present battery of tests. This situation would be unsatisfactory in the case of a nonpathogenic organism, but disastrous if the undetected virus were pathogenic for man.'[92]

The fact that monkeys can harbour such dangerous viruses should be a powerful argument against their use both in medical research and vaccine production.

Recently scientists have suggested that the African green monkey may be the origin of the human epidemic of AIDS.[93] It is thought that a similar virus, isolated from African green monkeys, infected humans and then mutated to the HIV virus that causes AIDS.[93] The monkey virus has actually been found in people who, like the African green monkey, remain unaffected. And there are even fears that a virus originally found in monkey kidney tissue may be linked to multiple sclerosis.[103]

Safety considerations and the inevitable shortage of monkeys caused by the rapidly expanding vaccine industry, finally rekindled interest in the use of human cells to produce polio vaccine and in 1961 Hayflick and Moorhead developed the use of human diploid cells.[91] Even so their introduction was delayed because it was felt that propagation of the cells might lead to abnormal and therefore dangerous properties. The idea was based on comparison with mouse cells, which react differently to human cells and do, indeed, become unstable when cultured in the test tube.[89] Once again animal-based research had proved misleading.

Today many viral vaccines such as polio, rubella, rabies, measles and smallpox can all be produced from human cells.[89] In Britain, Sabin vaccine is now prepared from human diploid cells,[94] but most of the polio vaccine used throughout the world is still produced from kidney tissue derived from African green monkeys and in some countries from rhesus monkeys.[95] And, although the killed Salk vaccine is traditionally made from kidney tissue, recent results show that this too can be produced from human cells.[96]

Animals are still used for testing vaccines but, once again, more sensitive and reliable *in vitro* techniques such as tissue culture are gradually replacing them. For instance, cell culture can be used instead of mice to test the potency of yellow fever vaccine[97] whilst a simple skin organ culture can be used instead of animals to ensure that cells used in vaccine production do not cause cancer.[98] According to the *Lancet*, 'In recent years many animal tests for the safety of viral vaccines have been replaced by cell culture tests which are more sensitive and reliable.'[99] Since *in vitro* systems are so enormously promising, the lack of an animal model of AIDS (see page 269) should not be seen as a problem in developing a vaccine but rather as an *opportunity* to develop more efficient techniques.

However, progress in eliminating the use of animals can be painfully slow, with experiments often repeated by government laboratories. Tests for identifying contaminants of polio vaccine have traditionally used small laboratory animals and, according to Dr Perkins of the World Health Organisation:

> 'I have not heard of any batch of vaccine that has been shown to be unsatisfactory purely on the grounds of tests in small laboratory animals. In 1962 there were 23 laboratories producing oral poliomyelitis vaccine; today there are 11 laboratories involved. If we take as an average 15 laboratories each year for the last 20 years, and if each produced 10 batches of vaccine a year, then 120,000 mice, 60,000 suckling mice, 30,000 guinea pigs and 60,000 rabbits have been used without adding anything to the safety of vaccines. At the very least there should be no necessity to repeat these tests by the control authority.'[100]

In 1982 the WHO finally recommended that such tests were unnecessary when human cells are used to produce the vaccine. And it is only recently that the requirement to test killed (Salk) polio vaccine on live monkeys, for residual live virus, has been dropped,[97] despite the fact that cell culture alternatives have long been far more sensitive.[100]

Other biological products such as hormones and anti-tumour antibiotics can be measured more accurately by high performance liquid chromatography (hplc), a sensitive nonanimal technique that divides a substance into its constituent parts for precise analysis. One example is the hormone oxytocin, where hplc has replaced a crude milk ejection test in female rats.[101] And recently, developments in hplc have made it possible to measure insulin purely by nonanimal means.[99] Hplc has also replaced the LD50 test for measuring the concentration of antitumour antibiotics such as dactinomycin.[102]

Compared with vital public health measures, vaccines had little impact on the dramatic decline in deaths over the past century. Even so, methods of production were extremely crude and often dangerous. The trend towards more sophisticated and reliable techniques does at least prove what antivivisectionists have always said – that there just *had* to be better ways for science to develop if only the obsession with animal experiments could be avoided. But, in some cases, the alternative is even simpler – not to do the research at all. And perhaps the strangest and most disturbing area of all is the use of animals in psychological and behavioural research, where the experiments tell us more about the scientists themselves than about human psychology.

1 Statistics produced annually by the Home Office. In *Hansard* (24 October, 1978), figures are given from 1878 to 1961
2 A. Rechtschaffen, et al, *Science*, 182-184, 22 July, 1983
3 T. H. Shepard, *Catalogue of Teratogenic Agents* (Johns Hopkins Press, 1976)
4 J. C. Fentress, *Science*, 704-705, 16 February, 1973
5 R. E. Brown and D. J. McFarland, *Animal Behaviour*, 887-896, volume 27, 1979
6 M. Beddow Bayly, *Clinical Medical Discoveries* (NAVS, 1961)
7 W. W. Holland and A. H. Wainright in *Epidemiologic Reviews*, volume 1, P. E. Sartwell (Ed.) (Johns Hopkins, 1979)
8 J. Stamler in *Cerebral Vascular Diseases* (2nd Conference), C. H. Millikan (Ed.) (Grune & Stratton, 1958)
9 See Chapter 2
10 W. P. U. Jackson and A. I. Vinik, *Diabetes Mellitus* (Edward Arnold, 1977)

11 R. Levine in an introduction to reference 12
12 B. W. Volk and K. F. Wellman, *Diabetic Pancreas* (Bailliere Tindall, 1977)
13 C. Singer and E. A. Underwood, *A Short History of Medicine* (Clarendon Press, 1962); see also reference 10
14 L. Stevenson, *Sir Frederick Banting* (Heinemann, 1947)
15 Diabetes Epidemiology Research International, *BMJ*, 479-481, 22 August, 1987
16 In 1934, a decade after the introduction of insulin, the *BMJ* (175, 28 July), observed that '. . . diabetic mortality is increasing all over the civilized world.' In 1983 (1855-1857, 11 June) the *BMJ* noted that the prevalence of diabetes melitus was increasing in many countries, including the USA, Finland, Israel, and other parts of Europe. In the UK the prevalence of diabetes is doubling every decade.
17 Recorded in R. Drewett and W. Kani's contribution to *Animals in Research*, D. Sperlinger (Ed.) (Wiley, 1981)
18 R. Drewett, *ibid*
19 W. A. R. Thomson, *Black's Medical Dictionary*, 33rd edition (A. & C. Black, 1981)
20 R. Brain, *Lancet*, 575-581, 17 October, 1959
21 G. Jefferson, *Lancet*, 59-61, 8 January, 1955
22 J. W. Lash and L. Saxen, *Nature*, 634-635, 27 August, 1971
23 *MRC Annual Report*, 1983-84
24 See, for example, the use of positron emission tomography in clinical studies of Parkinson's disease (D. B. Calne, et al, *Nature*, 246-248, volume 317, 1985) and stroke (R. J. S. Wise, et al, *Brain*, 197-222, volume 106, 1983)
25 See Chapter 5
26 K. Walker, *The Story of Medicine* (Hutchinson, 1954); see also reference 27
27 A. Hurst, *Lancet*, 950, 23 October, 1937
28 M. Anniko, *Laryng.-Rhinol*, 304, volume 60, 1981 (English abstract).
29 J. C. Petricciani, et al, *Investigational New Drugs*, 297-302, volume 1, 1983
30 P. P. Dendy and R. A. Meldrum in *Placenta: A Neglected Experimental Animal*, P. Beaconsfield and C. Villee (Eds) (Pergamon Press, 1979)
31 In its list of contents this journal indicates whether research has used human material. In an analysis by the writer during 1985, successive issues revealed that most basic cancer research uses animals or their tissues.
32 V. A. Lucas, Ph.D. Thesis, University of East Anglia, 1984, and M. Wilderholt and J. Kana, paper presented at International Symposium on the Eye (Berlin) 1984
33 D. F. Hawkins, *Lord Dowding Fund Bulletin*, 12-16, no. 23, 1985
34 T. Powell, et al, *BMJ*, 1013-1014, 17 October, 1981
35 T. Koppanyi and M. A. Avery, *Clinical Pharmacology & Therapeutics*, 250-270, volume 7, 1966, together with reference 40
36 J. F. Cavalla (Ed.), *Risk-Benefit Analysis in Drug Research* (MTP Press, 1981)
37 WHO list of essential drugs reproduced in *Bitter Pills* by D. Melrose (Oxfam, 1982)

38 *Extra Pharmacopoeia*, 28th edition, revised by J. Reynolds (Pharmaceutical Press, 1982)

39 *BMJ*, 525, volume 4, 1974

40 J. T. Litchfield in *Drugs in our Society*, P. Talalay (Ed.) (Johns Hopkins, 1964)

41 B. Reines, *Cancer Research on Animals* (NAVS, Chicago, 1986)

42 I. D. J. Bross, 'How we lost the war against cancer', *Congressional Testimony*, 1981, reproduced in reference 41

43 A. D. Dayan in reference 36

44 F. Gross (Ed.), *Antihypertensive Agents* (Springer-Verlag, 1977)

45 A. M. Breckenridge in reference 36

46 W. B. Pratt and R. W. Ruddon, *The Anticancer Drugs* (Oxford University Press, 1979)

47 F. I. McMahon, *Medical World News*, 168, volume 6, 1965

48 N. R. Farnsworth and J. M. Pezzuto, paper presented at University of Panama workshop sponsored by International Foundation for Science, 1982. Reproduced in *Lord Dowding Fund Bulletin*, 26-34, no. 21, 1984

49 For instance, S. E. Salmon in *Cloning of Human Tumour Stem Cells*, 291-312, (Alan Liss, 1980)

50 G. E. Foley and S. S. Epstein in *Advances in Chemotherapy*, vol. 1, A. Goldin and F. Hawking (Eds) (Academic Press, 1964)

51 *SCRIP*, 30, 16 January, 1987

52 G. Thompson, *Lord Dowding Fund Bulletin*, 25-28, no. 20, 1983

53 S. M. Hammond and P. A. Lambert, *Antibiotics and Antimicrobial Action* (Edward Arnold, 1981)

54 R. A. Bucknall in *The Use of Alternatives in Drug Research*, E. N. Rowan and C. J. Stratmann (Eds) (Macmillan, 1980)

55 H. F. Schaeffer, et al, *Nature*, 583-585, 13 April, 1978

56 D. H. Percy, et al, *British Journal of Experimental Pathology*, 41-49, volume 65, 1984

57 *Science*, 1355-1358, 20 December, 1985

58 G. Richards, *Quantum Pharmacology* (Butterworth, 1980)

59 See Chapter 3

60 *Lancet*, 887, 22 April, 1972

61 *New Scientist*, 218, 17 July, 1980

62 W. H. Inman in *Monitoring for Drug Safety*, W. H. Inman (Ed.) (MTP Press, 1980)

63 M. C. Henman and G. D. H. Leach, *British Journal of Pharmacology*, 591P, volume 80, 1983

64 R. Brent in *Advances in Experimental Medicine & Biology*, volume 27, M. A. Klingberg, A. Abramovici and J. Chemke (Eds) (Plenum Publishing, 1972)

65 *European Journal of Clinical Pharmacology*, 233-278, volume 11, 1977

66 K. E. Enslein, et al, *Benchmark Papers in Toxicology*, volume 1 (Princeton Scientific Publishers, 1983)

67 Health Designs Inc., *Toxicology Newsletter*, no. 3, 1984

68 R. Marks in *Animals and Alternatives in Toxicity Testing*, M. Balls (Ed.), et al (Academic Press, 1983)

69 J. J. Kaufman, et al, *Drug Metabolism Reviews*, 527-556, volume 15, 1984

70 *Food & Chemical Toxicology*, no. 2, volume 23, 1985

71 B. Ekwall, *Toxicology Letters*, 309-317, volume 5, 1980 *Toxicology*, 127-142, volume 17, 1980

72 G. M. L. Gyte and J. R. B. Williams, *ATLA*, 38-47, volume 13, 1985

73 P. D. Williams, et al, *In Vitro Toxicology*, no. 1, 23-32, volume 1, 1986-87

74 F. J. De Serres and J. Ashby (Eds), 'Evaluation of Short-Term Tests for Carcinogens' in *Progress in Mutation Research* (Elsevier Science Publishing, 1981); *Short-Term Tests for Chemical Carcinogens* H. F. Stich and R. H. C. San (Eds) (Springer-Verlag, 1981)

75 D. Paterson in *Animals in Research*, D. Sperlinger (Ed.) (Wiley, 1981)

76 Biovideograph's use in replacing cats for drug-induced effects on blood pressure, recorded in *Proceedings of the British Pharmacological Society*, 313P, 1-3 April 1981

77 Application for equipment grant for session 1982-1983 by University of Leeds Department of Physiology (17 March, 1983)

78 J. R. Walker, *Lord Dowding Fund Bulletin*, 6-9, no. 20, 1983

79 C. J. Dickinson, et al, *ATLA*, 107-116, volume 13, 1985

80 P. Townsend, Lord Dowding Fund Annual Lecture, Royal Society of Medicine, 13 November, 1985; see also *LDF Bulletin*, 6-8, no. 23, 1985

81 Reported in *BMJ*, 365, 30 July, 1983

82 *BMJ*, 996, 10 October, 1981

83 M. A. Hattwick and M. B. Gregg in *The Natural History of Rabies*, volume 2, G. M. Baer (Ed.) (Academic Press, 1985); see also *Lancet*, 628-629, 11 November, 1944

84 DHSS Memorandum on Rabies, 1977 (HMSO)

85 G. M. Baer (Ed.), *The Natural History of Rabies*, volume 2 (Academic Press, 1985)

86 *Nature*, 418, 1 February, 1969

87 K. G. Nicholson, et al, *Lancet*, 915, 24 October, 1981. Using human cell derived rabies vaccine the authors write: 'Thus, almost a century after the postexposure treatment of man began, effective prophylaxis appears to have been achieved.'

88 H. J. Parish, *Victory with Vaccines* (Churchill Livingstone, 1968)

89 L. Hayflick, *Laboratory Practice*, 58-62, volume 19, 1970

90 A. Lecornu and A. N. Rowan in *ATLA Abstracts*, Collected Reviews 1976-1979, M. Burkett (Ed.) (FRAME, 1980)

91 L. Hayflick, *Science*, 813-814, 19 May, 1972

92 P. B. Stones, *Laboratory Practice*, 40-44, volume 19, 1970

93 *Science*, 1141, 6 December, 1985

94 Letter dated 2 September, 1982 from A. J. Beale, Director of Biological Products, Wellcome Research Laboratories, to Robert Sharpe

95 A. J. Zuckerman, *BMJ*, 158, 18 January, 1986

96 B. Larsson and J. Litwin, *Developments in Biological Standardisation*, 241-247, volume 46, 1980

97 G. Langley, *New Scientist*, 12-16, 3 May, 1984

98 P. D. Noguchi, et al, *Science*, 980-983, 3 March, 1978

99 *Lancet*, 900-902, 19 October, 1985

100 F. T. Perkins, *Developments in Biological Standardisation*, 3-13, volume 46, 1980

101 R. A. Pask-Hughes, et al, *Analytical Proceedings*, 247-249, volume 18, 1981

102 In 1977 the US Food & Drug Administration (FDA) introduced regulations requiring LD50s to measure the potency of three antitumour antibiotics. Late in 1984 the FDA replaced this requirement with hplc. In the UK up until 1985 the *Biologicals Compendium* specified an LD50 for one of these drugs, dactinomycin, but have now dropped all mention of the LD50 (*Hansard* 11 July, 1985)

103 K. K. A. Goswami, et al, *Nature*, 244-247, 21 May, 1987

CHAPTER 7

Origins of madness

'Countless animals have been surgically dismembered, drugged, starved, fatigued, frozen, electrically shocked . . . maddened and killed in the belief that their behaviour, closely observed, would cast light on the nature of human kind.'

(Don Bannister,[1] Clinical Psychologist)

In 1942 C. S. Hall and S. J. Klein published an article on aggression in rats and remarked that '. . . the analysis of aggressive behaviour has recently become a popular subject for speculation and research.'[2] And so it proved. In the years ahead scientists would devise ever more ingenious methods of inducing aggression in their animal victims. The most popular technique turned out to be electric shock-induced fighting,[3] much used by Dr Roger Ulrich, one of the leading figures in aggression research. Since 1962 Ulrich's work at the University of Western Michigan has consisted largely in causing pain to animals and observing the resulting aggressive behaviour.[4]

A recent analysis of Ulrich's work showed how powerful electric shocks were administered through an electrified grid floor as a means of inducing fighting behaviour.[4] Shock intensity ranged up to the very painful 5 milliamps, which was often high enough to cause paralysis. A more sensitive strain of rat could not stand even half this intensity and several animals died during the experiment. In one study a pair of rats received no less than 15,000 shocks in a period of seven and a half hours. In another case five rats were shocked every day for 80 days causing them to fight '. . . more viciously, often cutting and bruising each other severely.'[4] Not content with painful electric shocks, Ulrich tried other distressing stimuli. The cage floor was heated, causing the rats to jump about, licking their feet as it grew hotter. Intense noise or cold was tried as were the effects of castration. Finally one pair had their whiskers cut off and were blinded by removal of their eyes.[4]

Eventually Ulrich had a change of heart. Writing in *Monitor*, the journal of the American Psychological Association, he explained his views:

'Initially my research was prompted by the desire to understand and help solve the problem of human aggression but I later discovered that the results of my work did not seem to justify its continuance. Instead I began to wonder if perhaps financial rewards, professional prestige, the opportunity to travel etc., were the maintaining factors and if we of the scientific community (supported by our bureaucratic and legislative system) were actually a part of the problem. Consider this; the withholding or removal of food or other reinforcers produces aggression in lab animals. Thus, when one considers that the United States with less than five per cent of the world's population consumes 45 per cent of the world's energy and resources, we become the cause of the anger and aggression in those people who don't have enough. 'When I finished my dissertation on pain produced aggression, my Mennonite mother asked me what it was about. When I told her she replied, "Well, *we* knew that. Dad always told us to stay away from animals in pain because they are more likely to attack." Today, I look back with love and respect on all my animal friends from rats to monkeys who were submitted to years of torture so that like my mother I can say, "Well, we know that."'[5]

Unfortunately Ulrich's 'years of torture' were not the isolated exploits of a solitary scientist. A review of the scientific literature between 1975 and 1978 by Washington's Animal Welfare Institute indicates that Ulrich was but one of many American investigators in aggression research.[3] Nor are these distressing experiments confined to the United States. The Postgraduate School of Psychology at the University of Bradford has carried out electric shock-induced fighting experiments in which animals were exposed to 60 electric shocks of 2 milliamp intensity over a 10 minute period.[6]

At University College Swansea the aptly named Dr Brain and his colleagues chose different models of aggression. In one case young mice were kept completely isolated for four weeks to induce aggression.[7] Dr Brain and his colleagues proved even more ingenious in 1983 when they published an article entitled 'Studies on

Tube Restraint-Induced Attack on a Metal Target by Laboratory Mice.' The article was published in *Behavioural Processes* and described how mice, tightly restrained in a perspex tube, attack a metal object dangled in front of them.[8] The effects of sex,

Experiments by technique used and use of anaesthesia (Great Britain 1986)

Number of experiments

Technique used		Use of anaesthesia			Total
		No anaesthesia [a]	Anaesthesia for part of experiment	Anaesthesia for whole experiment	
Other interferences with any of the special senses, or the brain centres	*for behavioural studies*	989	449	–	1,438
controlling them:	*for other purposes*	311	4,300	1,427	6,038
Interference with the central nervous system (other than centres controlling the special senses):	*for behavioural studies*	2,854[b]	26,253	–	29,107
	for other purposes	10,133[b]	23,339	12,491	45,963
Use of aversive stimuli, electrical or other:	*for behavioural training*	35,674	177	–	35,851
	for inducing a state of psychological stress, integral to experiment	3,524	8	–	3,532
Induction by any other means of a state of psychological stress integral to the experiment		22,126	369	–	22,495

[a] *Includes those experiments in which the subject of the study is the anaesthetic agent itself.*
[b] *In these experiments the interference with the central nervous system was minimal and was of a type for which anaesthesia was inappropriate and could have been harmful.*

Source: Statistics of experiments on living animals, 1986, Home Office (HMSO, 1987)

housing conditions, reproductive experience and castration were all examined. It was found that this model of aggression bears little resemblance to other models such as electric shock-induced fighting. So, more than 40 years after Hall and Klein's observations in 1942, aggression research is still thriving.

Sadly, aggression tests are but one area where painful procedures are used to investigate behaviour. In fact, animals are kept isolated from their companions or deprived of their mothers whilst others are starved, drugged and brain damaged, all to assess the resulting changes in behaviour. The vast majority of these experiments are carried out by experimental psychologists at university departments of psychology. But medical dictionaries define psychology as the study of the mind – of perception, thought, emotion, learning and behaviour, so human volunteers would seem the natural choice for research and observation. Yet, according to a survey by the British Psychology Society,[9] a wide variety of species are subjected to behavioural experiments including monkeys, cats, rabbits, rats, mice, pigeons, chicks and fish. The survey also revealed that surgery, drugs, electric shocks and food and water deprivation, are all commonly used during behavioural research.

Psychologists have been particularly interested in the effects of deliberately induced stress, such as that arising from social or maternal deprivation. Despite the distressing and well-known effects of early separation of mother and infant, experimental psychologists have worked hard to reproduce the same miserable condition in laboratory animals. Once again some of the worst examples come from the United States and Harry Harlow's work, but the UK has not been immune.

In an article entitled 'Effects of Various Types of Separation Experience on Rhesus Monkeys Five Months Later', Cambridge scientists Hinde, Leighton-Shapiro and McGinnis subjected infant monkeys and their mothers to various deprivation states.[10] In one case both mother and infant were removed from the colony and separated from each other. The tests reached some breathtaking conclusions: for instance the results are said to support the view that separation between mother and infant can, but need not, have long-term consequences. Furthermore, it is noted that those infants most affected by separation remained most sensitive a year later. In other recent experiments at Cambridge kittens were separated from their mothers to see the effect on play behaviour.[11] During the course of their observations, published in a report entitled 'Separation from the Mother and the Development of Play in Cats', the scientists found that separated kittens cried more than those still with their mothers: '. . . this call is seemingly given in distress.'[11]

At the University of Stirling tests have been carried out to see if the distress caused by solitary confinement could be reduced by allowing the animals to see their reflection in a mirror. In this case infant stumptail monkeys were used.[12]

Social deprivation is also used to investigate the value of play. For most people the idea that play has an important role in developing skills and experience, especially in children and young animals, is just common sense. Depriving young and normally sociable animals of companionship is therefore bound to be detrimental. But scientists are not so easily convinced. According to the popular science magazine *New Scientist*, Britain's Dorothy Einon has subjected young mice and rats to social isolation in order to assess the importance of play behaviour. She concluded:

> 'When we have sorted out play in the rat, so that we know whether any particular aspect is needed to protect against the ill-effects of isolation, I would like to look at other aspects of play in other animals. Different species play differently, and I am confident that play is used for a variety of different purposes in development.'[13]

Scientists not only induce fear and stress to observe the overall effects on behaviour, but to measure the physiological and biochemical effects. A recent issue of *Brain Research* described how UK researchers M. H. Joseph and G. A. Kennett measured the release of a brain chemical 5-HT during deliberately induced stress caused by forced immobilization of rats.[14] The animals were tied to a grid and electrodes in their brains recorded the biochemical effects of the stress. The authors claim their findings are relevant to naturally arising stress disorders in people, although it is not clear how such experiments solve the problems that lead to stress in the first place.

Another example, this time at the Welsh National School of Medicine in Cardiff, subjected rats to electric shocks to see the effects of acute physical stress on plasma concentrations of a substance called α-melanotropin.[15]

Despite more than half the world being permanently

undernourished, experiments to discover the specific effects of
food deprivation on sexual and aggressive behaviour and on the
behaviour of infant animals, have featured prominently in
experimental psychology. A whole research programme at the
University of Manchester is devoted to the behavioural and
developmental effects resulting when baby rats are starved. J.
Tonkiss and J. L. Smart of the University's Medical School note
that,

> 'The field of research on the behavioural effects of early life
> undernutrition is strewn with equivocal and contradictory
> findings. It has been suggested that some of this confusion
> may have arisen from the use of different strains of rats in
> different laboratories.'[16]

Then they proceeded to test the behaviour of two different
strains of rat, born to mothers starved during gestation and
suckling. Manchester's Professor John Dobbing had already
shown that, '. . . rats starved during the suckling period are
stunted in their growth and their brains are also permanently
stunted.'[17] This resulted in impaired learning and changes in
physical activity. But, in referring to Dobbing's research into the
development of the brain, which it partly funded, the Medical
Research Council warned that,

> '. . . differences in behaviour are not necessarily a direct
> consequence of impaired growth of the brain; the vulnerable
> period is also one when the baby shows the greatest emotional
> dependence on its mother. The only way of separating the
> possible causes is to bring up baby rats without their mothers,
> and substantial progress has been made in rearing rats
> artificially. An answer should soon be available as to whether
> there are behavioural effects of malnutrition which are not
> related simply to the presence of the mother but to intake of
> food.'[17]

Has the Medical Research Council ever stopped to consider
how starving children would react if they realized that precious
resources are being allocated to food deprivation tests like those
at Manchester? For whatever the results, their desperate need is
for food and improved social and environmental conditions.
Deliberately starving animals will not help them.

As the British Psychological Society report revealed, electric
shocks are commonly used in experimental psychology. We have
seen already how shocks are used to induce stress or to make
animals fight each other but, as a form of punishment or

conditioning, they are also employed to investigate the learning process. For instance the *Quarterly Journal of Experimental Psychology* describes how Cambridge scientists administered electric shocks to the eyes of rabbits to make them blink.[18] Shortly before the shocks the animals were shown a light that they came to associate with the pain to their eyes. Eventually they became conditioned to blink with the light alone. The use of eye shocks as a method of conditioning has been described by scientists at the Medical Research Council's Unit on Neural Mechanisms of Behaviour as '. . . useful for analysis of vertebrate learning.'[19]

It might be suggested that tests like these, to develop the underlying theories of learning, have been necessary for the development of behaviour therapy, to treat phobias and other neurotic disorders. But as psychologists Drewett and Kani have explained,

'. . . it would be curious to argue that it would have been *impossible* to carry out the relevant research on human subjects; for how could a therapeutic method for use with human beings be based on principles of learning which could be demonstrated and investigated in dogs and rats, but which could not be demonstrated and investigated in human volunteers? Indeed, one could argue that the development of behavioural therapy might have been more rapid if more of the relevant research had been carried out on human volunteers rather than on animals (for instance, the importance of imagery would probably have been defined earlier).'[20]

Then we come to the endless round of brain-damage tests in which animals are deliberately injured to observe the effects on behaviour. Traditionally doctors have used clinical observation (and post-mortem examination) of patients with brain injuries to relate changes in behaviour to specific parts of the brain (see also Chapter 6). For example, it was found that bilateral removal of the mesial temporal lobes in people resulted in profound amnesia, indicating that the hippocampus plays an important role in human memory.[21] But once again this is not sufficient for the vivisectors and there is a vast literature covering the effects of brain damage on the behaviour of laboratory animals.

At the Institute of Psychiatry in London, 6 to 12-day-old monkeys were subjected to brain damage to see how this affected their sense of touch and accuracy in reaching for objects. The report, published in *Brain Research*,[22] explained how similar tests with juvenile monkeys had already been carried out but, the authors argued, nothing is known about the effects on very young animals. Recent reports from the Universities of Oxford and Reading show how cynomolgus and rhesus monkeys were subjected to brain damage to observe the effects on learning and memory.[23] Similar tests have been carried out on pigeons and rhesus monkeys at Newcastle[24] and Cambridge.[25]

Brain damage tests are also carried out to assess the effects of an enriched environment on recovery. A recent edition of *Developmental Psychology* carries a report by Manchester scientists H. B. Katz and C. A. Davies in which baby rats were subjected to brain damage caused by food deprivation.[26] One-month-old animals were only half their normal weights. They were then housed in boring or interesting conditions to assess the effects of environment on recovery. Research by Dorothy Einon and her colleagues at Durham[27] and similar experiments at University College Swansea investigated the role of social isolation in recovery from brain damage.[28] Not surprisingly animals living in a social and enriched environment fared better than those kept in solitary confinement or in boring conditions.

Psychologists also study the effect of drugs on animal behaviour and this includes drug addiction and withdrawal. Considerable effort is directed towards already known mood-changing drugs such as stimulants and minor tranquillizers. For instance, doctors have known about the ill-effects of amphetamine for years yet researchers continue to reproduce symptoms in laboratory animals. In 1983 scientists at the Clinical Research Centre in Harrow administered increasing doses of amphetamine to vervet monkeys for 35 days to observe the effects on behaviour and chemical changes in the brain.[29] From day seven their behaviour started to deteriorate. They suffered tremors, showed poor balance when disturbed and indulged in self-destructive grooming. After day 29 they became grossly over-reactive to external events. Finally they were killed and their brains examined for biochemical changes. In the same year the journal *Psychopharmacology* carried another report by scientists at the Clinical Research Centre in which marmosets had a tube surgically implanted into their brains so that amphetamine could be administered and the effects on behaviour observed.[30] And despite the huge literature on Librium and Valium, animal tests gave no warning of the terrible social problems that followed their widespread use: of one million people given minor tranquillizers every year in the UK, 100,000 become drug dependent.[31] Yet, in 1984 Malcolm Lader of London's Institute of Psychiatry wrote that, 'Dependence to benzodiazepines [for example Librium and Valium] is difficult to induce in animals.'[32]

Sadly there are all too many human addicts without subjecting animals to further cruelties. Counselling, group therapy and community care are surely the best approaches to treatment and care, since animal models are unlikely to shed light on the complex psychological facets of human drug addiction and withdrawal. The view that animal experiments have little to offer seems to be shared by the *British Journal of Addiction*, which concentrates overwhelmingly on the patient. In a survey of papers published during 1980 and 1981 only one of the 75 articles specifically dealt with animal experiments, in this case a report from Switzerland on the effects of nicotine on rodent behaviour.[33] And in the case of alcoholism, the Alcohol Studies Centre in Scotland has stated that '. . . nothing of clinical relevance has been achieved to date from the vast range of animal experiments' and that '. . . the animal models of addiction . . . are not relevant to human addiction.'[34]

In recent years the use of animals in behavioural experiments like those described here, has been increasingly condemned not least by psychologists themselves. One of the fiercest critics is Alice Heim, Fellow of the British Psychological Society and past President of the Psychology Section, British Association for the Advancement of Science.

'But for me the big questions are: have we learned much that is new or beneficial from these thousands of experiments? And – whatever may have been learned – is the infliction of so much pain and terror warrantable? And, finally, is there any possible justification for duplicating and reduplicating this sort of experiment or variants of it, when the results are known and are readily ascertainable by means of films, books and articles in Journals?'[35]

In August 1985 on behalf of Mobilization for Laboratory Animals, a coalition of leading antivivisection societies in the UK, the writer compiled a dossier listing 32 recently published examples in which animals had been subjected to stress, starved, shocked, drugged and brain-damaged during the course of behavioural research. The report showed that most of the experiments were funded by the taxpayer through one or other of the Research Councils and produced results that appeared both trivial and obvious. The results of social isolation, the need for an enriched environment in recovery from brain damage, and the effects of food deprivation early in life, are all examples where the outcome seems only too obvious. Mobilization sent copies to the Home Office, arguing that the use of animals in painful behavioural experiments should no longer be licensed under new legislation, but the government refused even to acknowledge the report, which eventually had to be raised in a parliamentary question.[36]

Later on, when Harry Cohen, MP, tabled an amendment to the Animals (Scientific Procedures) Bill prohibiting the use of animals in behavioural tests, Home Office Minister David Mellor reiterated the usual defence – that the experiments are necessary for research into mental illness and neurological disorders.[37] Psychological disturbances and mental illness are hard enough to define in people let alone animals so it is difficult to take Mellor's claim seriously.

In his *Diseases of Civilization* (Paladin Books, Granada, 1981), Brian Inglis describes an experiment undertaken by Stanford University psychologist David Rosenham that shows just how difficult it is to diagnose mental disorders in human patients.[38] Rosenham and seven friends presented themselves individually at 12 different mental hospitals in the United States. None had any psychiatric problems but they agreed to claim only that they occasionally heard voices saying words that sounded like 'empty', 'hollow' and 'thud'. Otherwise they were to behave normally and stick to the truth when questioned. The outcome was astonishing: all were admitted to hospital, seven as schizophrenics, one as a manic depressive. They were prescribed huge quantities of drugs. Neither psychiatrists or hospital staff had any suspicions and they were finally released from hospital up to 52 days after admission! If doctors find mental illness so difficult to diagnose in humans, how can such conditions be reproduced in animals?

In 1954 the *British Medical Journal* remarked that, 'Few neurological and probably no psychiatric disorders can be adequately reproduced in animals.'[39] But the vivisectors, despite considerable scepticism and even ridicule,[40] would not give up so easily. And soon there would be a more profitable incentive as psychotropic or mood-changing drugs – antidepressants, sedatives, stimulants, sleeping pills and tranquillizers – became big business. By the 1980s 15-20 per cent of all prescriptions were for psychotropic drugs[31] with the minor tranquillizer Valium alone enjoying world sales of £250 million a year.[41] Furthermore, a report published in 1969 by the Office of Health Economics, an organization set up by the UK's drug industry, predicted that by 1990 new drugs intended primarily for social purposes would be in use, so that nearly every individual would be taking mood-changing drugs either continuously or at intervals.[42] The drugs would be 'pacifiers' and 'general comforters'.

Companies with a newly patented psychotropic medicine could therefore expect huge profits even if they only managed a small slice of the market. Not surprisingly animal models became very popular. By 1983 the British National Formulary listed 18 antidepressant drugs available for treatment whilst the December 1985 issue of *Psychology in Practice* revealed that some 108 antidepressants were currently under development. A survey of new drugs introduced onto the world market between 1975 and 1984 listed 21 new products to treat anxiety, in addition to those currently used.[43] And by 1983 between 20 and 30 major tranquillizers of the phenothiazine class were available to British doctors.[44]

Animal models of anxiety are created by subjecting animals to unpleasant conditions such as electric shocks or by dosing them with chemicals known to produce anxiety in people and presumed to do so in animals.[45] Development of new major tranquillizers, to treat serious mental illness such as schizophrenia, mania, dementia and personality and behaviour disorders, does not depend on animal models of the actual disease because these do not exist. Instead, drug development depends on animal models of the *side-effects* of already known drugs![46] The idea is based on the assumption that specific side-effects often parallel a drug's useful effects. So if the test drug produces the same side-effects as already established medicines, it is presumed to have the same beneficial effects. Many of the side-effects of major tranquillizers, including drug-induced parkinsonism, tardive dyskinesia, catalepsy and suppression of vomiting, are all caused by effects on a specific aspect of brain chemistry. New drugs are therefore tested on animals to see if they affect the same part of the brain. For instance, rats are

DRINKING MONITOR WITH ELECTRICAL PUNISHMENT FOR TESTING ANTI-ANXIETY DRUGS (ANXIO-METER MODEL 102)

In testing the efficiency of tranquilizers use in suppressing anxiety, rats are first deprived of water for 48 hours. The animals are then allowed to drink from the water dispenser which is programmed to deliver an electrical punishment each time twenty licks are completed. Due to this punishment, the animals develop anxiety and refrain from further drinking.

A number of pharmacological substances are able to suppress this anxiety allowing the thirsty rats to drink to their satisfaction, in spite of the punishment.

The efficiency of anti-anxiety drugs can be measured by comparing the drinking behavior of punished drugged rats to punished non-drugged rats.

Columbus Instruments has developed commercially an instrument described by R. A. Vogel in 1971 for the purpose of testing anxiolytic substances. In the mean time R. A. Vogel's test has become the standard in psychopharmacology, but until now researchers have had to build their own equipment. "ANXIO-METER" is equipped with two electronic counters, one for counting number of licks on the water spout, second for counting the number of electric shocks delivered to the animal through the same spout. In general, rats treated with highly efficient drugs are punished more often and number of electrical shocks they absorb during drinking is a direct measure of particular drug efficiency.

"ANXIO-METER" is also equipped with an A.C. current shock generator and a session timer which terminates the experiment after a preset time interval as described by Vogel. Other time intervals are available upon request.

"ANXIO-METER" is the first commercially available instrument of its kind and is intended for pharmacologists and biochemists developing or testing new sedatives and tranquilizers at universities, government research institutes and pharmacological industry.

ORDERING INFORMATION:
Catalog #117 — Complete single animal Anxio-Meter Model 102, including shocker but excluding animal cage and water dispenser.
Catalog #118 — Complete 81 x (6) animals "Anxio-Meter" model 102-6 including shocker but excluding animal cages and water dispensers
Catalfog #118-1 — Animal cage equipped with stainless grid floor and water dispenser to be used with single animal or six animal "Anxio-Meters"

From Columbus Instruments Catalogue, 1984

monitored for the onset of catalepsy after being injected with the test drug. Or vomiting is induced in dogs and they are then watched to see if the new drug suppresses the effect. Unfortunately, when the tests prove successful, they almost inevitably lead to drugs with serious built-in side-effects such as catalepsy, parkinsonism and tardive dyskinesia, where patients lose control of their muscles. As prescriptions for tranquillizing drugs rose by 76 per cent between 1961 and 1975, there was a corresponding rise of 114 per cent in drugs to treat parkinsonism over the same period.[47] And in a recent report from Edinburgh's City Hospital, over 50 per cent of new cases of parkinsonism were thought to be drug-induced.[48]

Because the animal models are so poor, the beneficial effects of many drugs used to treat mental disorders have first been discovered through clinical observation of patients rather than by animal experiments. The tranquillizing action of chlorpromazine (Largactil) and its ability to treat schizophrenia were first observed in patients.[49] In fact, '... the most famous psychotropic drugs', writes Turan Itil of the New York Medical College's Department of Psychiatry, 'were discovered by chance.'[50]

But it is the animal models of depression that have the most appalling history. A recent editorial in *Psychological Medicine*[40] credits Harry Harlow, Professor of Psychology at the University of Wisconsin, with much of the current interest in primate models of depression. In 1962 Harlow's experiments on the effect of separating infant rhesus monkeys from their mothers were published in the *Journal of Child Psychology and Psychiatry* (B. Seay, E. W. Hanson and H. F. Harlow, 123-132, volume 3, 1962).

When separated, the infants became very upset and showed extreme agitation followed, a few days later, by withdrawal, lethargy and depression. These symptoms had *already* been described in human infants and children separated from their parents[40] and, because doctors referred to them as depressed, Harlow claimed his separated infant monkeys were likewise 'depressed'. Harlow's experiments were subsequently repeated by Hinde in Cambridge and Kaufman and Rosenblum in New York.[40] In 1965 Harlow described his work over the previous decade:

'For the past ten years we have studied the effects of partial social isolation by raising monkeys from birth onwards in bare wire cages ... These monkeys suffer total maternal deprivation ... More recently we have initiated a series of studies on the effects of *total* social isolation by rearing monkeys from a few hours after birth until 3, 6 or 12 months of age in a stainless steel chamber. During the prescribed sentence in this apparatus the monkey has no contact with any animal, human or sub-human.'[51]

As a result Harlow was able to show that,

> ". . . sufficiently severe and enduring early isolation reduces these animals to a social-emotional level in which the primary social responsiveness is fear.'[51]

A later article describes how Harlow and his colleague Stephen Suomi tried to induce depression by allowing baby monkeys to cling to cloth surrogate mothers who would then turn into monsters.

> 'The first of these monsters was a cloth monkey mother who, upon schedule or demand, would eject high-pressure compressed air. It would blow the animal's skin practically off its body. What did the baby monkey do? It simply clung tighter and tighter to the mother, because a frightened infant clings to its mother at all costs. We did not achieve any psychopathology. However, we did not give up. We built another surrogate monster mother that would rock so violently that the baby's head and teeth would rattle. All the baby did was cling tighter and tighter to the surrogate. The third monster we built had an embedded wire frame within its body which would spring forward and eject the infant from its ventral surface. The infant would subsequently pick itself off the floor, wait for the frame to return into the cloth body, and then cling again to the surrogate. Finally, we built our porcupine mother. On command, this mother would eject sharp brass spikes over all of the ventral surface of its body. Although the infants were distressed by these pointed rebuffs, they simply waited until the spikes receded and then returned and clung to the mother.'[51]

Another ingenious device invented by Harlow and Suomi was the 'well of despair'. Because depression in humans has often been described as a state of helplessness and hopelessness, as if in a well of despair, they built a vertical chamber with stainless steel sides sloping inwards to form a rounded bottom, and placed young monkeys in it for periods up to 45 days.[51] After a while the animals would sit huddled in a corner of the chamber. The confinement apparently produced '. . . severe and persistent psychopathological behaviour of a depressive nature.'[51]

Harry Harlow died in 1981 and the author of the editorial in *Psychological Medicine* who referred to his '. . . genius, imagination and forethought'[40] was none other than Stephen Suomi, Harlow's partner in some of the most sickening experiments ever recorded. Such is Harlow's influence that, even now, maternal and social deprivation experiments are still reported in the scientific literature. A survey by Dr Martin Stephens found that between 1961 and 1984 368 papers were published on maternal deprivation.[52]

Since Harlow's 'imaginative' research, many more animal models of depression have appeared, so have patients really benefited? A recent report from the Lafayette Clinic in Michigan notes that '. . . it is easy to understand why many clinicians and investigators have been extremely sceptical of alleged animal models of depression',[53] and goes on to describe the '. . . poor track record of most if not all animal models to date in accurately predicting clinically effective antidepressants.'

> 'These observations are highlighted by the fact that almost every significant advance in antidepressant drug treatment from the discovery of iproniazid and imipramine to the recently introduced "second generation" class of antidepressants has resulted either from astute clinical observation or serendipity.'[53]

Imipramine's beneficial effects were first discovered in clinical trials with depressed patients but then scientists accidentally found that the drug reversed chemically-induced hypothermia in mice.[54] Using this as an animal model of depression, about a dozen more drugs very similar to imipramine were developed. But the model proved unreliable because other drugs, such as iprindole, mianserin, salbutamol and beta-flupenthixol, whose antidepressant properties were also first discovered following their clinical evaluation in depressed patients, were all predicted to fail by the mouse hypothermia test.[54] Although commonly used, this dubious test can only be expected to provide yet more drugs similar in action to imipramine. And what possible connection can it have to the problems that cause people to be depressed in the first place?

Other animal models include mouse-killing behaviour in rats, and a technique known as 'learned helplessness', also proposed as an animal model of anxiety. In this case animals are subjected to unavoidable electric shocks but, later on, when escape is made possible, they fail to learn how to avoid the shocks. Their helplessness has been 'learned' from the first part of the experiment. This achievement represents the latest animal model of depression and anxiety.[55]

However they are tested, the overall impact of psychotropic

drugs is difficult to assess because, although they have strong advocates there are serious drawbacks. The major tranquillizers such as chlorpromazine, used to treat serious mental illness, are usually credited with reducing the population of mental hospitals. An alternative view of their action is the 'chemical cosh' such is their tremendous power of sedation in dampening emotions. And although the number of occupied beds in mental hospitals has indeed declined, the number of attendances by outpatients has increased considerably, suggesting that the overall level of illness may be much the same. In fact the number of *new* patients at UK mental hospitals has remained virtually unchanged since 1971.[56]

Nevertheless the original credit for the 'open-door system' in which mentally ill patients are, whenever possible, returned to the community, does not go to the major tranquillizing drugs or to electroconvulsive therapy, but to a French doctor, Philippe Pinel (1745-1826).[57]

For centuries the mentally ill were brutally treated, incarcerated and punished as if they were possessed by demons. If the credit for bringing an end to this brutal period is to be accorded to any one man it must be to Pinel who was physician to the Biûtre Prison in Paris where large numbers of the mentally ill were incarcerated. Pinel rejected the medieval idea of possession, believing instead that the mentally ill were sick people requiring treatment rather than punishment.

His aim was to try and build up the patient's confidence by taking him seriously and trying to understand his problems. He freed 'prisoners' rather than keeping them confined and, although they still remained ill, they ceased to be rebellious and disorderly because they were treated kindly.

Unfortunately, although the major tranquillizers often help to keep difficult patients under control and are no doubt popular with hospital authorities, they fail to tackle the underlying causes of the disease. On the other hand psychotherapy, for instance involving psychoanalysis as developed by Freud, although taking longer, can locate the cause and actually provide a cure. A Harley Street specialist observes that little attention is given to drug-free psychotherapy in the treatment of schizophrenia, such is the obsession with modern medicines. He states that,

'... the overuse of drugs has helped swell the numbers of so-called chronic psychiatric patients. It is much easier to prescribe drugs than to try and find out why a patient has developed to the point when the label "schizophrenia" has to be given. This easy way has made many psychiatrists blind or unwilling to seek methods which would change the patient into a normal and productive human being.'[58]

Much the same is true of anxious and depressed patients: continual use of minor tranquillizers and antidepressants diverts attention from the underlying causes and only masks the original problem. And with the benzodiazepine minor tranquillizers, if the original cause is not identified and cleared up, the patient may simply go on taking the tablets, eventually becoming addicted to the treatment.

Ironically, withdrawal symptoms often include intense anxiety, the very condition the drugs were initially meant to cure. Major tranquillizers create their own problems too. They are a major cause of parkinsonism and a related group of symptoms called tardive dyskinesias where patients lose control of their muscles. As we have seen, these side-effects are often an inevitable consequence of the way the drugs are developed. Tardive dyskinesia (t.d.) starts with involuntary movement of the tongue and facial muscles, but in more extreme cases the arms and legs jerk uncontrollably. In some cases the effects are irreversible. According to Dr David Hill, a clinical psychologist at Walton Hospital in Chesterfield, about 38 million people have t.d. worldwide with more than 25 million rendered permanently unable to control the muscles in their tongues or, in many cases, their entire bodies.[59]

Although the drug manufacturing industry employs a wide variety of animal models to produce new psychotropic medicines, the vast majority of behavioural experiments reported in the scientific literature have nothing to do with drug development at all and are carried out by university departments of psychology. No one can reasonably claim that they have had the slightest impact on the overall level of mental illness, which has remained virtually unchanged in recent years. We can only agree with clinical psychologist Don Bannister, a member of the Medical Research Council's External Scientific Staff at High Roads Hospital in Ilkley when he concludes that, '. . . the sad reflection must be that the countless animals who have died in psychological experiments have not only died cruelly, but in vain.'[1]

1 D. Bannister in *Animals in Research*, D. Sperlinger (Wiley, 1981)
2 C. S. Hall and S. J. Klein, *Journal of Comparative Psychology*, 371-383, volume 33, 1942
3 J. Diner, 'Physical and Mental Suffering of Experimental Animals – A Review of Scientific Literature 1975-1978' (Animal Welfare Institute, Washington, 1979)
4 D. Pratt, *Painful Experiments on Animals* (Argus Archives, 1976)
5 R. Ulrich, *Monitor*, Journal of the American Psychological Association, 1978
6 For example, R. J. Rodgers and A. Depaulis, *Pharmacology Biochemistry & Behaviour*, 451-456, volume 17, 1982; R. J. Rodgers and R. M. J. Deacon, *Physiology & Behaviour*, 183-187, volume 26, 1981
7 S. Al-Malike and P. F. Brain, *Animal Behaviour*, 562-566, volume 27, 1979
8 P. Brain, et al, *Behavioural Processes*, 277-287, volume 8, 1983
9 British Psychological Society: Scientific Affairs Board, *Bulletin of the British Psychological Society*, 44-52, volume 32, 1979
10 R. A. Hinde, et al, *Journal of Child Psychology & Psychiatry*, 199-211, volume 19, 1978
11 P. Bateson and M. Young, *Animal Behaviour*, 173-180, volume 29, 1981
12 J. R. Anderson, *Animal Learning & Behaviour*, 139-143, volume 11, 1983
13 D. Einon, *New Scientist*, 934-936, 20 March, 1980
14 M. H. Joseph and G. A. Kennett, *Brain Research*, 251-257, volume 270, 1983
15 J. F. Wilson and M. A. Morgan, *Psychopharmacology*, 67-72, volume 68, 1980
16 J. Tonkiss and J. L. Smart, *Developmental Psychobiology*, 287-301, volume 16, 1983
17 MRC Annual Report, 1981-82 (MRC)
18 J. M. Pearce, et al, *Quarterly Journal of Experimental Psychology*, 45-61, volume 33B, 1981
19 M. Glickstein, et al, *Proceedings of the Physiological Society*, 30-31P, March, 1983
20 R. Drewett and W. Kani in *Animals in Research*, D. Sperlinger, (Wiley, 1981)
21 J. O'Keefe and D. H. Conway, *Physiological Psychology*, 229-238, volume 8, 1980
22 J. V. Brown, et al, *Brain Research*, 67-79, volume 267, 1983
23 D. Gaffan, et al, *Quarterly Journal of Experimental Psychology*, 173-222, volume 36, 1984
24 A. Sahgal, *Behavioural Brain Research*, 47-58, volume 11, 1984
25 A. Sahgal, et al, *Behavioural Brain Research*, 361-373, volume 8, 1983
26 H. B. Katz and C. A. Davies, *Developmental Psychobiology*, 47-58, volume 16, 1983
27 D. F. Einon, et al, *Quarterly Journal of Experimental Psychology*, 137-148, volume 32, 1980
28 J. C. Dalrymple-Alford and D. Benton, *Behavioural Neuroscience*, 23-34, volume 98, 1984
29 R. M. Ridley, et al, *Neuropharmacology*, 551-554, volume 22, 1983
30 L. E. Annett, et al, *Psychopharmacology*, 18-23, volume 81, 1983

31 MRC Annual Report, 1982-83 (MRC)
32 M. Lader, *Progress in Neuro-Psychopharmacology & Biological Psychiatry*, 85-95, volume 8, 1984
33 J. Schlatter and K. Bättig, *British Journal of Addiction*, 199-209, volume 76, 1981
34 Letter dated 15 January, 1985 from the Director, Alcohol Studies Centre, Paisley, Scotland, in reply to personal communication
35 A. Heim, speech at NAVS Annual Public Meeting, 1980 (published by NAVS)
36 *Hansard*, 173-174, 18 December, 1985
37 Third sitting of House of Commons Standing Committee on the Animals (Scientific Procedures) Bill, 6 March, 1986
38 B. Inglis, *Diseases of Civilization* (Paladin Books, Granada, 1981)
39 *BMJ*, 1364, 12 June, 1954
40 S. J. Suomi, *Psychological Medicine*, 465-468, volume 13, 1983
41 *SCRIP*, 27, 17 June, 1985
42 Medicines in the 1990s, a technological forecast, Office of Health Economics, 1969
43 *SCRIP*, 20-21, 23 December, 1985
44 P. Parish, *Medicines: A Guide for Everybody* (Penguin, 1983)
45 H. Lal and M. W. Emmett-Oglesby, *Neuropharmacology*, 1423-1441, volume 22, 1983
46 P. Worms and K. G. Lloyd, *Pharmacology & Therapeutics*, 445-450, volume 5, 1979
47 A. Melville and C. Johnson, *Cured to Death* (New English Library, 1983)
48 P. J. Stephen and J. Williamson, *Lancet*, 1082-1083, 10 November, 1984
49 'Tranquillizing action', J. T. Litchfield in *Drugs in Our Society*, P. Talalay (Ed.) (Oxford University Press, 1964); 'Schizophrenia', A. M. Breckenridge in *Risk – Benefit Analysis in Drug Research*, J. F. Cavalla (Ed.) (MTP Press, 1981)
50 T. M. Itil, *Progress in Neurobiology*, 185-249, volume 20, 1983
51 Recorded in P. Singer's book *Animal Liberation* (Thorsons Publishing Group, 1983) and references therein
52 Maternal Deprivation Experiments in Psychology, M. L. Stevens, 1986 (American Antivivisection Society, National Antivivisection Society, Chicago and New England Antivivisection Society)
53 N. Sitaram and S. Gershon, *Progress in Neuro-Psychopharmacology & Biological Psychiatry*, 227-228, volume 7, 1983
54 B. E. Leonard, *ibid*, 97-108, volume 8, 1984
55 S. F. Maier, *Psychopharmacology Bulletin*, 531-536, volume 19, 1983
56 *Social Trends*, no. 16, 1986
57 K. Walker, *The Story of Medicine* (Hutchinson, 1954)
58 J. Bierer, *Nature*, 468, 6 October, 1983
59 *The Guardian*, 16 July, 1985

CHAPTER 8

No peace for animals

'The more I look back I see their greatest fear is in people finding out how the animals are treated and thereby initiating steps to correct that.'

(Donald Barnes,
Brookes Airforce Base, Texas[1])

On January 11, 1980 Donald Barnes was dismissed from his position as behavioural psychologist at the Brookes Airforce Base in San Antonio. For 15 years he had worked in the Weapons Effects Branch where rhesus monkeys and baboons are given powerful electric shocks to force them to carry out tasks of human design. When fully trained they would be irradiated or exposed to nerve gases and their antidotes to see just how much their performance was impaired. In a statement[1] Barnes explains how he became more and more unwilling to expose these creatures to pain and, therefore, less productive in the number of 'subjects' trained and utilized in the experiments. He describes how,

> 'In years past, I was ordered to keep a death watch on these irradiated subjects, which meant, simply, to see what happened until they died of radiation injury. Do you have any idea how miserable it is to die from radiation injury? I do, I've seen so many monkeys go through it.'[1]

As his anxiety at the plight of irradiated monkeys increased, Barnes started to protest on their behalf, stressing the futility of many experiments. Inevitably his days at the Base were numbered and when Shirley McGreal of the International Primate Protection League, paid a visit during 1979, it became only too clear that he was no longer trusted. Dr McGreal was carefully steered away from the building in which Barnes worked.[2]

In November 1980, I recorded an interview with Donald Barnes in a Los Angeles hotel room where he produced photographic evidence to support his statements. He described how rhesus monkeys are restrained in a specially made and expensive primate equilibrium platform (PEP) to mimic pilots exposed to radiation caused by the bomb they have just dropped.[3] Electric shocks are used to train the monkey to 'fly'

PATHOLOGY REC $\;$ ST AND REPORT OF ANIMAL OR TISSU(			DATE OF SUBMISSION 30 Apr 79
ANIMAL SPECIES AND ID NUMBER M. mulatta 254D		SUBMITTED BY *(Investigator/Office Symbol/Extension)* Dr. Farrer/RZW/3684	
SEX Male	AGE	WEIGHT *(Kg)*	JOB ORDER WORK UNIT NUMBER 775705E1 (77578537)
DATE AND HOUR OF DEATH *(Approximate if exact is not known)* 30 Apr 79		PRELIMINARY PHONE REPORT OF FINDINGS DESIRED ☐ YES ☐ NO	
FOR VSP USE ONLY			
ACCESSION NUMBER N79-241		DATE OF REPORT 3 Dec 79	BACT	PHOTOS
OTHER PERTINENT DATA *(Provide as much of the history of the animal's use or illness as possible. A reproduced copy of the animal's records will assist the pathologist in his examination and evaluation of findings.)*			

History:
Exposed to 360 Rads Gamma (Co^{60}) on 27 Apr 79.

Gross:
Raised, pale circumscribed areas were seen on the surface of the lung. They were 5-10 mm in diameter and on cut surface were hollow. These lesions were considered to be bronchiectatic areas due to P. simicola.

Micro:
Small intestine, skin, aorta, stomach, kidney, liver, heart, cerebrum, cerebellum, eye, esophagus, tongue, gall bladder, epididymis, urinary bladder, pancreas, and peripheral nerve - No lesions seen.
Lymphoid tissue - Lymphoid tissue throughout the body had a "washed out" appearance with loss of germinal centers and decreased numbers of lymphocytes in adjacent areas. Reticulo-endothelial hyperplasia was seen in germinal centers and in medullary areas. Mild hemorrhage was seen.
Bone marrow - Fewer cells and less immature cells were seen in the bone marrow.
Testicle - The spermatic tubular epithelium was vacuolated and reduced numbers of spermatogonia and spermatocytes were seen. No sperm were seen.
Lung - Large peribronchiolar areas with extensive neutrophil, macrophage, and plasma cell collections were seen. Mite cross sections were seen in the centers of many of these areas. Dark brown granules were seen in macrophages associated with these areas and were considered to be mite pigment granules.

Diagnoses: 1. Radiation sickness, acute, diffuse, moderate, lymph nodes, spleen, small intestine, testicle, bone marrow, colon, pancreas, Rhesus, etiology radiation.
2. Pulmonary acariasis, chronic, moderate, lung, etiology P. simicola.

SIGNED

GENE B. HUBBARD, Maj, USAF, VC   Cy: VSR file
Comparative Pathology Branch
Veterinary Sciences Division

the platform in such a way as to maintain a relatively horizontal position, that is between ±10°: more than that and the monkey receives an electric shock through the footplate. The animal would then be irradiated to see how his performance is affected. According to a memo dated November 7, 1979 and sent to a Mr A. Rahe, Barnes notes that in one experiment, the animals received 100 rads immediately upon initiation of the PEP. One hour later, they received 35 rads and an hour after that, a further 105 rads delivered over 15 minutes. 'At 7½ hours into the "mission", these subjects received a massive dose of 1,200 rads delivered at 12r/min. In total, therefore, each subject received 1,440 rads; for most purposes, however, the 1,200 rad dose was the important one for performance effects.' But the experiments turn out to have little value except, as Barnes himself notes, 'to generate more worthless experiments.'[1] A report by the School of Aerospace Medicine summarizing data over six years from 210 rhesus monkeys concludes that, 'The effects of such factors

PATHOLOGY REG :T AND REPORT OF ANIMAL OR TISSU		DATE OF SUBMISSION	
ANIMAL SPECIES AND ID NUMBER Rhesus monkey #182A	SUBMITTED BY (Investigating Office Symbol/Extension) RRW/Dr. Farrer		
SEX Male	AGE	WEIGHT (Kg)	JOB ORDER-WORK UNIT NUMBER 77570539
DATE AND HOUR OF DEATH (Approximate if exact is not known) 5 Jun 78	PRELIMINARY PHONE REPORT OF FINDINGS DESIRED ☐ YES ☐ NO		
FOR VSP USE ONLY			
ACCESSION NUMBER N783339	DATE OF REPORT 19 Jul 78	BACT	PHOTOS
OTHER PERTINENT DATA (Provide as much of the history of the animal's use or illness as possible. A reproduced copy of the animal's records will assist the pathologist in his examination and evaluation of findings.)			

Clinical Data:
Died in chair during experimental procedure. Animal was thought to have been overdosed with atropine.

Gross description:
The only lesion noted was a slight reddening of the lungs.

Histologic description:
Lung – Small areas of hemorrhage were noted throughout the lung sections.
CNS – There was a minimal amount of extravasation of erythrocytes around blood vessels in the brain.
Salivary gland, lymph node, large intestine, eyes, and kidney – No lesion recognized.

    Diagnosis: Hemorrhage, focal, minimal, lung and brain, etiology undetermined.

SIGNED
ROBERT E. SCHMIDT, Lt Col, USAF, VC   Cy: VSR file
Chief, Comparative Pathology Branch
Veterinary Sciences Division

as dose rate, fasting time, radiation quality and performance tasks on the radiation emesis syndrome are still undetermined.'[4]

Furthermore it is no easy matter to train a monkey to carry out such tasks and whilst 3-5 milliamps may be sufficient to *maintain* performance, much higher levels are required to train the animal in the first place:

'The shock generators are designed and manufactured by BRS [Behavioural Research Systems] and deliver at least 50mA at 1,200 volts. I couldn't even guess at the number of times I've seen these units used at full power to punish a slow learner or to otherwise "reinforce" undesirable behaviour – well into the thousands – however, the learning process is replete with other dangers for the monkey as frustration leads to other self-destructive behaviours, e.g., biting hunks of meat from an arm or hand, pulling out hair until the subject is bald in accessible spots.'[1]

Even the restraint devices themselves cause suffering:

'As the animal struggles to free itself, it often loses its teeth to the neckbar, gains severe abrasions on the abdomen (often wearing entirely through the abdominal wall), or so severely chafes its ankles that they bleed and become infected: and the animal is shocked and shocked again (sometimes hundreds and hundreds of times per day), until it either does the experimenter's bidding or is "flunked out" to another programme requiring no training such as laser beams in the macula of the eye or centrifuge work at g-forces which are permanently damaging.'[1]

Other experiments carried out at the Base include the use of dogs to investigate vomiting following irradiation.[3,5] For instance, experimental veterinarian Lieutenant Colonel Joel Mattsson tested atropine and dexamethazone for their possible value as anti-emetic drugs. According to Mattsson's monthly activities report for April 1980, in which he describes 'significant achievements':

'Twenty-three more dogs were tested in the other eight treatment groups this month. Results continue to indicate a large inter-individual variability in sensitivity to radiation. This probably also exists in humans, accounting for much of the confusion and contradictory results in the literature on drug effectiveness for preventing radiation emesis.'

Animals surviving these experiments are used in other studies.

Mattsson states that dogs used in radiation-induced sickness tests were subsequently used for training in battlefield medicine.[5] And after sublethal irradiation studies, rhesus monkeys are '. . . turned in for use in other experiments', according to the Animal Statement. This also describes how animals exposed to 360 rads, are expected to experience radiation symptoms such as nausea, vomiting, anorexia and fatigue. Treatment is not recommended however, because it '. . . could confound the performance results and invalidate the data.' Donald Barnes gives a rare and horrifying glimpse of the undercover world of weapons research, inevitably the most secret use of animals.

In the UK the name that has become synonymous with chemical and biological warfare is Porton Down. During the First World War a new dimension in warfare was heralded by the gas clouds drifting over the trenches. They inflicted agonizing injuries and deaths upon thousands of troops and in 1916, in hurried response to the first German attack with chlorine gas, the British government opened Porton Down in Wiltshire, later to be known as the Chemical Defence Establishment.

In their book *A Higher Form of Killing : The Secret Story of Gas and Germ Warfare*,[6] Robert Harris and Jeremy Paxman piece together startling evidence that has only recently come to light. They describe how Porton Down scientists then, as now, made extensive use of human volunteers and how they established a farm in 1917 to breed the animals needed for top secret experiments. Reports of tests made in those early years have now been released to historians and make horrifying reading:

> 'They give some idea of the scale and substance of the grim research which has made Porton a top target for anti-vivisectionists. Cats, dogs, monkeys, baboons, goats, sheep, guinea pigs, rabbits, rats and mice were variously tethered and caged outdoors in the trench system and indoors in the gas chambers for exposure to gas clouds. Chemicals were squirted into their faces and injected into them, and bullets, sprays and bombs fired into, over and at them. With the discovery of mustard gas, bellies and backs were shaved and the chemical rubbed in; some animals were opened up and their organs smeared with mustard, the wound then stitched back together and the symptoms which developed noted. The Establishment became such a prominent centre of vivisection

that it later developed its own strain of 'Porton mice', now a standard laboratory ánimal in use throughout the world.'[6]

J. B. S. Haldane records that the physiologists at Porton Down '. . . had considerable difficulty in working with a good many soldiers because the latter objected so strongly to experiments on animals, and did not conceal their contempt for the people who performed them.'[7] Between 1952 and 1970 Porton Down's Microbiological Research Centre consumed over 1,000 monkeys, nearly 200,000 guinea pigs and 1,250,000 mice.[6]

In more recent times, scientific reports, together with Ministry of Defence disclosures, reveal that Porton Down's Chemical Defence Laboratories have tested riot control gas on animals,[8] exposed monkeys and other animals to nerve gas,[9] subjected dogs to the poisonous effects of hydrogen cyanide,[10] tested the effects of rubber and plastic bullets on live sheep[11] and shot monkeys through the head with ballbearings to investigate the effects of high velocity missiles.[12] Animals used include sheep, pigs, dogs, rabbits, guinea pigs, mice, rats and monkeys.[13] From 1981 to 1983 40,100 animal experiments were '. . . notified to the Home Office'[14] whilst over the past five years 131 rhesus monkeys have been used in the chemical defence programme.[15] In 1984 the number of experiments rose by 1,400 to 10,900 – a necessary increase claims the Ministry of Defence, because of '. . . the need to ensure that effective protection and treatment remain available to our forces to meet continuing developments in the threat posed by the Soviet Union's chemical warfare capability.'[16]

Like the Brookes Airforce Base, Porton Down scientists have been testing possible antidotes to nerve gases like soman, sarin, tabun and XV.[17] In a letter to Lieutenant Colonel John Howell, Assistant Executive Secretary of USAF's Scientific Advisory Board, Professor E. F. Domino reports on his trip to the UK, Germany, Sweden, Norway and the Netherlands when he held detailed discussions on chemical warfare.[18] Professor Domino visited Farnborough and Porton Down and discussed various antidotes to soman. He refers to the UK's 'vigorous human testing programme' and writes: 'The British use of human volunteers is really a very important supplement to their excellent animal research programme.'[18] According to the Ministry of Defence, about 200 service personnel take part in experiments connected with the chemical defence programme every year.[14] All participants, the MoD states, are volunteers, fully informed of the dangers but Porton Down's need for human guinea pigs shows once again that animal tests can never be relied upon. For instance, in the search for antidotes to soman, substances that proved effective in rats and mice, failed when tested in guinea pigs and monkeys.[19]

Other details of Porton Down's work with nerve gases emerge through occasional scientific articles. A recent report in the *Journal of Pharmacy and Pharmacology* describes how rhesus monkeys were exposed to soman during tests for possible antidotes.[20] After a medium dose of soman, together with antidote, the animals became prostrate with violent convulsions. After 1-11 minutes they lost consciousness and breathing became slow and laboured and the animals appeared very close to death. After a high dose of soman the animals collapsed within a minute, with violent convulsions and laboured breathing. The animals surviving this stage regained consciousness after 10-30 minutes: 'The animals then made attempts to crawl about the cage but relapsed after about an hour and died.'[20]

In 1983 Porton Down scientists published experiments with CS riot control gas.[8] Guinea pigs, rats and mice were forced to inhale the gas for one hour a day, five days a week for up to 120 days. Many of the animals died during this period: 'During the first month of the experiment 46% of the high dose group of guinea pigs perished.'[8] The tests were supposedly carried out to assess toxicity yet CS had already been in use for many years.[6] CS had originally been developed by Porton Down scientists as a more effective agent than CN and was tested by the British when faced by rioters in Cyprus during 1958. Victims felt their eyes burn and water, their skin itch and their noses run whilst at the same time they coughed and vomited between gasps for breath. Later on CS was used in Northern Ireland and by the Americans in Vietnam.

Porton Down's work has not been restricted to chemical weapons. In 1942 British work on a biological bomb led to researches on the remote island of Gruinard off the north west coast of Scotland. On that occasion an anthrax bomb was released over sheep specially gathered by Porton Down scientists.[6] A day after the bomb exploded, the sheep began to die. At the end of each round of tests the sheep were dragged to the edge of some nearby cliffs and flung over. The hilltops were then exploded to cover the carcasses. A major scare occurred when one of the carcasses was blown clear and actually drifted over to the mainland, causing an outbreak of anthrax in cattle. Today Gruinard is utterly abandoned and no one is allowed to live there. Dramatic warning signs are posted at 400 yard intervals along its beaches:

GRUINARD ISLAND
THIS ISLAND IS GOVERNMENT
PROPERTY UNDER EXPERIMENT.
THE GROUND IS CONTAMINATED
WITH ANTHRAX AND DANGEROUS.
LANDING IS PROHIBITED

Gruinard remains an embarrassing reminder of Porton Down's attempts to develop biological weapons but even after the war the UK decided to continue biological research, according to an investigation by the *Observer*.[21] Discoveries were passed to Canada and the United States who alone could afford to manufacture weapons. The three countries also carried out joint tests during 1952. In Operation Cauldron cages of monkeys were sprayed with plague bacillus during top secret tests off Cellar Head on the Hebridean island of Lewis. The *Observer* investigation revealed that during a joint conference in Canada in 1958, the United States chemical corps minuted '. . . it was agreed . . . studies should be continued on aerosols . . . all three countries should concentrate on the search for incapacitating and new-type lethal weapons.'[21]

Even today the UK still maintains an interest in germ warfare. In 1979 the Ministry of Defence recruited a dozen specialists to '. . . take care of critical defence problems in microbiology'[6] whilst government responses to parliamentary questions indicate that animals are indeed used to '. . . ensure protection against biological warfare.'[22] Porton Down also set up a respectable sounding Public Health Service Laboratory to investigate dangerous pathogens but suspicions increase when defence scientists appear as co-authors on published papers, for instance in the case of legionnaire's disease.[23]

Scientific reports also show how monkeys have been shot in the head in wound tests at Porton Down. In a paper entitled 'Experimental High-Velocity Missile Head Injury', authors Allan, Scott and Tanner describe how anaesthetized rhesus

MONTHLY ACTIVITIES REPORT

WEAPONS EFFECTS BRANCH
April 1980

Significant Achievements

1. **Antiemetics for Radiation Induced Sickness** (77570538)

   Two new treatment groups have been added to the study; atropine and dexamethasone. Preliminary results indicate neither drug has particular antiemetic value. Testing will continue, however, until sample sizes are large enough to minimize type II error. Twenty-three more dogs were tested in the other eight treatment groups this month. Results continue to indicate a large inter-individual variability in sensitivity to radiation. This probably also exists in humans, accounting for much of the confusion and contradictory results in the literature on drug effectiveness for preventing radiation emesis. In the dog study, the variability in the combined drug treatments is less than for controls or single drug treatments.

2. **Rhesus-Fascicularis Comparison Study** (77570539)

   A statistical analysis of the baselines of the twelve monkeys was completed. It was determined that ten were adequately trained; so, testing was begun. To date, four doses (0.079, 0.14, 0.25, 0.44 mg/kg) of atropine have been administered. The three highest doses produced mild to severe disruption of PEP performance. It is projected that the study will be completed by mid June.

3. **Battlefield Medicine** (ED93991X)

   Dr. Mattsson conducted six half-day laboratory training sessions on emergency airway and chest procedures. A group of veterinarians and physicians are being trained to be laboratory instructors for the battle-field medicine course that begins next September. In addition, five physicians from the RAM course were given laboratory training on performance of orotactile intubation, intracath endotracheal intubation, cricothyreotomy, tracheostomy, and relief of pneumothorax by needle thoracentesis and tube thoracostomy. Anesthetized dogs that had previously been subjects on a radiation study (antiemetic drug study) were used.

JOEL L. MATTSSON, LT COL, USAF
Acting Chief, Weapons Effects Branch

Evidence from Brookes Air Force Base

monkeys were shot just above the eye to investigate the effects of injury:

'In the 20 experimental animals a penetrating injury was inflicted by a steel ball with a diameter of 3.2mm fired at a range of 5 or 10 metres from a smooth barrel with an estimated impact velocity of 1,000 metres/sec.'[12]

The researchers then watched the animals to see how long it took them to die. The article in *Injury* records survival times ranging from 2 to 169 minutes.[12] The experiments sparked off a fierce parliamentary row and led to further revelations about

Porton Down's work. According to the Ministry of Defence, sheep, pigs, rabbits and monkeys have all been used in wounding experiments.[14] Over the period 1977-1983, '... wounding studies have involved 194 penetrating injuries and 438 non-penetrating or blast-type injuries.'[14] Government states that '... all animals in wounding studies are deeply anaesthetized when wounded' but some are allowed to recover from the anaesthetic and suffer the consequences:

> 'Most animals used in these studies died or are killed while still under anaesthesia but a proportion are allowed to recover so that the development of the injury may be studied.'[14]

In 1965 the Home Office's own internal report on the Cruelty to Animals Act 1876 (the Littlewood Committee) revealed that animals are subjected to the effects of blast or shock type injuries at Porton Down but, contrary to later claims by the MoD that all animals are anaesthetized when wounded,[24] the Committee were informed that no anaesthetics were used because they '... might produce minor degrees of change in tissue simulating the effect of blast damage.'

Despite widespread condemnation from politicians of all parties, government continues to defend this type of work, arguing that it contributes to the treatment of wounds for both service personnel and civilians. But human beings and animals are physiologically and anatomically different and, as we have already seen, surgeons have long argued that using animals to develop surgical techniques can be entirely misleading.[25] Furthermore doctors at the famous Royal Victoria Hospital in Belfast who have to deal with the real victims of a continuing war, believe the Porton Down experiments are of no value to them in treating human patients.[26] In any case it is well known that during times of war, surgical techniques develop rapidly because doctors must instantly apply their skills to the wide range of unexpected injuries that arise. They cannot afford to wait for the results from animal experiments, even assuming the results were relevant. Sir Cecil Wakeley, consulting surgeon to Kings College Hospital has noted how an almost uninterrupted succession of wars in recent centuries '... undoubtedly led to an appreciable increase in surgical knowledge, and ... many of the greatest surgeons gained their experience on the battlefield.'[27] One of these was military surgeon Ambroise Paré who strongly advised against the 'cruel' cautery to stop bleeding after amputation. His new method was to ligature the vessels so as to stem serious bleeding, although it was another 200 years before the method was accepted in England. Paré also abolished the use of boiling oil to cauterize gunshot wounds, using a bland dressing instead. Wakeley goes on to assess advances during the twentieth century and writes:

> 'The two world wars, especially the second, played an important part in the advance of surgery. The first world war did much to establish orthopaedic surgery as a speciality.'[27]

During the Second World War, surgery for wounds of the chest and heart became a relatively common procedure and many of the fundamental skills of heart surgery were developed.[28]

The inevitable conclusion is that wounding experiments are not carried out to advance the treatment of human injuries but to test the efficiency of weapons of war. And documents obtained by the *Mail on Sunday* indicate that numerous animal tests have been carried out at Porton Down to observe the injuries caused by different kinds of bullets.[29] These animals are not dying to save lives – the usual defence of animal tests – but so that people can be killed more effectively.

As Ann Clwyd MP has pointed out,[30] there is a peculiar obscenity in inflicting such pains on our fellow creatures in the science of destruction and killing. So it is not surprising that warfare tests have always presented a difficult public relations problem for the authorities. Fort Detrick, the United States' main centre for the study of biological weapons, and by 1960 the biggest user of guinea pigs in the world, sponsored a well-equipped scout pack, supplied the local paper with a weekly gossip column and made speakers available for local discussion groups.[6] Porton Down was more reserved but, like the British government, often relied on claims that the animals did not really suffer too much. Despite the evidence presented by Robert Harris and Jeremy Paxman in their book *A Higher Form of Killing* (Paladin Books, Granada, 1983), the government's Littlewood Committee reported in 1965 that,

> '... we were given categoric assurances that no experiment had been, or was being, performed at Porton that caused pain or acute distress to any animal for an appreciable time.'

More recently the Ministry of Defence assured Parliament that of those experiments '. . . connected with the chemical and biological defence programme, many involve minor, if any discomfort.'[31] But if tests with nerve and riot control gases only cause 'minor discomfort', why are these agents regarded as important weapons of war?

The Ministry of Defence also states that, '. . . if an animal is found to be suffering severe pain which is likely to endure, it is at once painlessly killed.'[31] This is one of the pain conditions attached to the Cruelty to Animals Act which has controlled animal experiments in Britain since 1876. But, since no one has ever defined 'severe' or 'likely to endure', the condition becomes utterly meaningless and is clearly used to give the impression that animals are protected against excessive levels of pain. Nevertheless published examples definitely show that animals do undergo prolonged suffering, as in the case of eye tests with tear gas. The scientific textbook *Current Approaches in Toxicology* reveals how Porton Down scientists Ballantyne and Swanston tested the peripheral sensory irritant CN (tear gas) in the eyes of conscious rabbits.[32] Some of the injuries, involving '. . . diffuse redness of the lids with moderate swelling', started to decline after three days but others gradually deteriorated so that after seven days the scientists observed '. . . gross opacification of the cornea with deformity and/or ulceration.' After a further seven days these injuries remained much the same with the animals obviously blinded by the tear gas. In another case rats and rabbits were forced to breathe highly toxic chemicals used as smoke screens in military training.[33] The chemicals used were already known to be highly dangerous both to animals and people but the Porton Down scientists observed their victims for up to 14 days after exposure. Most of the animals died before the experiment was complete and many must have suffered terribly as the fumes gradually destroyed the delicate lining of their throats. Not only that but, in some cases, their lungs became waterlogged and started to bleed. For one of the smoke mixtures five rabbits died within 24 hours with the remaining five surviving up to one week. For the other mixture only three out of ten rabbits were still alive after two weeks. The researchers were able to conclude that the '. . . high degree of local irritancy produced by the two materials concurred with previous observation on human exposures and dogs' although '. . . further work is being carried out using repeated exposures to hexachloroethene zinc chloride smokes which determine whether progressive lung damage occurs with recurrent exposure.'[33]

Would the suffering in experiments like these be described as severe and sufficiently prolonged to breach the government's pain condition? Or would it be dismissed as 'substantial' pain and therefore not applicable. In any case, despite assurances by the Ministry of Defence, such experiments may not even be subject to the Cruelty to Animals Act since the Crown is not bound by the Act.

Another useful ploy is to argue that detailed information about animal usage would be detrimental to national security. This was the Minister's response to a question by Roland Boyes MP who asked how many animals had been used in cyanide poisoning tests.[34] The Minister could not deny that animals had been used (we had copies of the scientific publications) but claimed that, for security reasons, no details could be given. Yet, earlier that same year, when questioned about CS riot control gas, he had given a detailed breakdown of species and the number of animals used.[35] Could it be that the animal most often used in cyanide experiments is the dog whilst with riot gas, it is rats, mice and guinea pigs – species for which the public has considerably less sympathy?

In fact the use of dogs in cyanide tests dates back to the early years of chemical warfare. Porton Down's first head of Physiology, Joseph Barcroft, wanted to settle a dispute between the British and French about the effectiveness of hydrogen

---

CS Gas (Animal Experiments)

Mr. Boyes asked the Secretary of State for Defence how many animals have been used in experiments testing CS gas for each year from 1974 to 1983 and 1984 to date; how many animals died as a result of the experiments or were killed at the conclusion of the experiments; what species were used; and if he will make a statement.

Mr. Lee [*pursuant to his reply*, 6 February 1984, c. 476]: The table as printed contained some errors. A corrected table is as follows:

Year	Species	Number used	Number exposed to CS	Died during experiment	Killed at conclusion
1974	Rats	100	100	61	39
1975	Rabbits	10	10	0	10
1976	Mice	300	*225	31	0
	Rats	200	*150	8	0
	Guinea Pigs	200	*150	39	0
1977	Mice	—	—	11	†258
	Rats	—	—	9	†183
	Guinea Pigs	—	—	16	†145

* Animals not exposed to CS were required as controls.
† 1977 figures refer to the continuing work initiated in 1976; no further animals were used.

Hansard, 15 April 1984

cyanide.[6] The French had tested the gas on dogs, all of which died and, as a result, believed it would make an effective chemical weapon. But British tests were carried out on goats, which survived. In order to settle the dispute Barcroft decided to test the gas himself and, without putting on a mask, stepped into a gas chamber with a 1 in 2,000 concentration of hydrogen cyanide. He also took a dog in with him:

> 'In about thirty seconds the dog began to get unsteady, and in fifty seconds it dropped on the floor and commenced the characteristic distressing respiration which heralds death from cyanide poisoning. One minute thirty five seconds after the commencement the animal's body was carried out, respiration having ceased and the dog being apparently dead. I then left the chamber. As regards the result upon myself, the only real effect was a momentary giddiness when I turned my head quickly.'[6]

Over 60 years later Porton Down was still carrying out cyanide experiments on dogs. In 1982 the scientific journal *Archives of Toxicology* described how beagle dogs were exposed to cyanide in order to test an antidote called DMAP, a substance already shown to be effective in treating cyanide poisoning in man.[10] The conscious animals were dosed with DMAP and then hydrogen cyanide to see if the antidote prolonged survival. Signs of cyanide intoxication were observed and all the animals lost consciousness for ten minutes. There were 'marked' signs of laboured breathing. Four out of six dogs recovered after one hour but one suffered two major epileptic fits. The sixth dog died 44 minutes after the injection of cyanide. Three other dogs were injected with cyanide alone and acted as untreated 'controls'. They suffered laboured breathing and spasms. Respiration and pulse ceased within one and a half minutes and the corneal reflex was lost. Despite all the *published* information, the Ministry of Defence could still not give details of cyanide tests because this would be '. . . detrimental to national security.'[34]

If all else fails there is still the good old standby – that military research is of a purely *defensive* nature. Yet, even without developing new weapons, it is obvious that defence research must entail a study of the potential *offensive* capacity of such agents, if only to enable better protection to be developed. This was admitted by the Holland Committee, set up by the British government after the First World War to decide on

Porton Down's future. They concluded,

> '. . . that it is impossible to divorce the study of defence against gas from the study of the use of gas as an offensive weapon, as the efficiency of the defence depends entirely on an accurate knowledge as to what progress is being made or is likely to be made in the offensive use of this weapon.'[36]

Successive generations of government ministers must have deeply regretted the Committee's honesty, because it blows their traditional line of defence against opponents of both animal experiments and chemical and biological weapons. In fact, the elaborate defensive measures necessary to protect against nerve gases like soman could never be made available to the general public, only for limited military use. So animal tests would provide absolutely no protection for civilians should there be a full scale chemical or biological war and people would die in agony just as the monkeys did in Porton Down's soman tests. The only sensible answer is to negotiate the weapons away.

If a full-scale chemical and biological arms race were to begin again – a definite possibility as more and more countries develop the necessary capability – recent advances in genetic engineering could make nerve gas and anthrax bombs look like crude remnants from the dark ages. A United States defence spokesman has claimed that genetic engineering could solve one of the major disadvantages of biological warfare, that it is limited to diseases that occur naturally somewhere in the world.

> 'Within the next 5 to 10 years, it would probably be possible to make a new infectious micro-organism which could differ in certain important respects from any known disease-causing organisms. Most important of these is that it might be refractory to the immunological and therapeutic processes upon which we depend to maintain our relative freedom from infectious disease.'[36].

Could it be, as former director of Berlin University's Institute of Biology, Professor Jacob Segal has suggested,[37] that AIDS does not originate from African Green monkeys at all, but was 'engineered' from two separate viruses at America's secret biological weapons centre in Fort Detrick? He argues that scientists injected the new virus into prisoners during the mid 1970s in return for their freedom, but were unaware of the real nature of their terrible creation. If this is true, the AIDS virus has a distinct advantage over the man-made super germ predicted by the American defence spokesman in 1969. AIDS is not only 'refractory' to the immune system, it actually destroys it.

1 Statement by D. Barnes reprinted in the International Primate Protection League's (IPPL) newsletter, March, 1980
2 IPPL newsletter, March, 1980
3 Reproduced in *Animals Defender*, January/February 1981 (Journal of the National Antivivisection Society)
4 M. G. Yochmowitz, J. L. Mattsson and V. L. Bewley, 'Radiation Emesis Repository (1971-77): Analysis' (Report SAM-TR-78-26, USAF School of Aerospace Medicine, Aerospace Medical Division (AFSC), Brookes Air Force Base, Texas 72835, September 1978)
5 J. L. Mattsson, Monthly Activities Report (April, 1980), Weapons Effects Branch, Brookes Air Force Base, Texas 72835
6 R. Harris and J. Paxman, *A Higher Form of Killing* (Paladin Books, Granada, 1983)
7 J. B. S. Haldane, reproduced in reference 6
8 T. C. Marrs, H. F. Colgrave, N. L. Cross, M. F. Gazzard and R. F. R. Brown, *Archives of Toxicology*, 183-198, volume 52, 1983
9 R. H. Inns and L. Leadbeater, *Journal of Pharmacy & Pharmacology*, 427-433, volume 35, 1983 and reference 20
10 T. C. Marrs, J. E. Bright and D. W. Swanston, *Archives of Toxicology*, 247-253, volume 51, 1982
11 Ministry of Defence, *Hansard*, 501, 21 February, 1984
12 I. V. Allen, R. Scott and J. A. Tanner, *Injury*, 183-193, volume 14, 1982
13 *Hansard*, 248, 15 February, 1984 and references 10 and 14
14 Ministry of Defence, *Hansard*, 570-572, 22 March, 1984
15 Ministry of Defence, *Hansard*, 617, 23 February, 1984
16 Ministry of Defence, *Hansard*, 266, 17 May, 1985
17 Scientific reports mention these gases, for example reference 19 describes antidote tests against four nerve gases – soman, sarin, tabun and XV
18 Letter from E. F. Domino dated 7 December, 1977
19 R. H. Inns and L. Leadbeater, *Journal of Pharmacy & Pharmacology*, 427-433, volume 35, 1983
20 P. Dirnhuber, M. C. French, D. M. Green, L. Leadbeater and J. A. Stratton, *Journal of Pharmacy & Pharmacology*, 295-299, volume 31, 1979
21 *Observer*, 21 July, 1985
22 In a statement defending the use of animals at Porton Down, the Ministry of Defence said that '. . . it is also necessary to ensure protection against biological weapons' (reference 14)
23 For example, in a paper on experimentally induced legionnaire's disease (*Journal of Pathology*, 349-362, volume 139, 1983), three authors are listed as being from Porton Down's Public Health Laboratory Service and one (M. Broster) from the Chemical Defence Establishment.
24 'All animals are deeply anaesthetized when wounded' according to Mr Lee, Ministry of Defence (*Hansard*, 570-572, 22 March, 1984)
25 In his book *The Futility of Experiments on Living Animals* (NAVS, 1962), Dr

Beddow Bayly quotes many surgeons who argue that developments in surgery must come from clinical work rather than animal experiments. See also Chapter 3

26 W. Rutherford (Senior accident and emergency consultant) Newsletter (Belfast) 21 February, 1984; also reference 29

27 C. Wakeley, *Lancet*, 903-906, 9 November, 1957

28 The Wellcome Museum of the History of Medicine (Science Museum, London, November, 1986)

29 *Mail on Sunday*, 19 February, 1984

30 Ann Clwyd, *Sanity*, February, 1986 (CND journal)

| 31 Ministry of Defence, *Hansard*, 125-126, 13 March, 1984

32 B. Ballantyne and D. W. Swanston in *Current Approaches in Toxicology*, B. Ballantyne (Ed.) (John Wright & Sons, 1977)

33 T. C. Marrs, W. E. Clifford and H. F. Colgrave, *Toxicology Letters*, 247-252, volume 19, 1983

34 Ministry of Defence, *Hansard*, 47-48, 3 December, 1984

35 Ministry of Defence, *Hansard*, 248, 15 February, 1984

36 Reproduced in reference 6

37 *Sunday Express*, 26 October, 1986

CHAPTER 9

The survival of vivisection

'Sadly, young doctors must say nothing, at least in public, about the abuse of laboratory animals, for fear of jeopardising their career prospects. In permitting this situation to continue the august bodies which administer the affairs of the [medical] profession do us no service. Animals should never be abused in the course of medical research, and a truly ethical profession should not allow this to continue.'

(Dr E. J. H. Moore, *Lancet*, 1986[1])

In May 1984 an American cell of the Animal Liberation Front 'visited' the notorious head injury laboratory at the University of Pennsylvania where primates are subjected to deliberate brain damage by forces of up to 3,000gs. The animals are fitted into an acceleration device so that their heads can be thrust forwards and sideways to simulate whiplash or non-impact brain injuries. ALF removed 60 hours of videotape which showed that sadism, unscientific practices and incompetence were the by-laws of the laboratory.[2] A half-hour digest of these tapes has been widely shown in the UK because Glasgow researchers Adams and Graham received the damaged brains for study at the Neuropathology Unit of the Southern General Hospital. It shows how baboons are no longer under anaesthesia at the time of injury but had to be tied down in order to restrain them. After the injuries, the dental enamel helmets put onto their heads were removed with a hammer and chisel, requiring hundreds of blows! Not only that but the brain-damaged animals were treated as objects of amusement and ridicule, sometimes being forced to watch as other baboons were being injured. During one 'fun' session, a researcher actually said 'Let's hope the AV people don't get hold of *this* tape!'[3]

Despite the sickening and irrefutable evidence, Glasgow's internal inquiry, eventually set up in January 1985 and chaired by Professor Bryan Jennett, is reported to have cleared the research, stating that animals were '. . . treated humanely before, during and after the experiment.'[3] Professor Jennett is a member of the Advisory Board of the American Head Injury Foundation, which, in November 1984, made an award to another Board member Dr Thomas Gennarelli, one of the principal investigators present during much of the video filming. Eventually the Glasgow scientists were forced to eat their words when, on July 18, the United States National Institutes of Health, which provided the $13 million for the project, announced an immediate suspension of all federal funding for the research. This followed a sit-in at its offices. The University of Pennsylvania then decided to 'indefinitely suspend' all primate research at their head injury laboratory because the case had '. . . shaken public confidence in its work and had harmed the reputation of the University as a whole.'[3]

We now begin to appreciate the desire for secrecy because, whatever the alleged benefits, people will not tolerate scenes of cruelty that rightly belong to a concentration camp. But the evidence presented here shows that vivisection is doubly absurd: not only does it cause suffering to animals, its overall contribution to our health is negligible. Since 1876 when the Cruelty to Animals Act was passed, over 170 million animal experiments have been carried out in the UK alone. Of these, 85 per cent were performed since 1950 yet, by then, the rapid fall in deaths that commenced more than a hundred years earlier, had largely levelled out. So the vast majority of experiments can have had little impact and today we know that society's control of the infectious epidemics rests primarily on efficient public health services and a good standard of living. The increase in life-expectancy, then, can be directly traced to these sources. Even now infectious diseases like TB and whooping cough are far more common amongst working-class people than in the professional workers of social class I, despite the availability of modern drugs and vaccines. And, despite the huge increase in animal experiments since 1950, our overall health could now be deteriorating.[4]

The conclusion is inevitable: when it comes to real advances in health, animal experiments are irrelevant. To make matters worse, vivisection has constantly proved misleading, diverting attention and resources from more reliable sources of information, much of which focuses on preventing disease rather than treating its symptoms. So it is not surprising that the really important advances have come not by experimenting on animals but through methods that directly relate to people.

Yet, despite the evidence, the vivisection industry has worked hard to maintain the illusion that all is well in the world's laboratories. Traditionally, as we have seen, there are two main lines of defence: firstly that animals do not really suffer at all or, if they do, it is nothing more than mild discomfort and, in any case, they are better looked after than most domestic pets. In the UK there is the added bonus that experiments are said to be 'stringently controlled' by the Home Office: in 1986 there were 18 inspectors to supervise 442 registered laboratories, 20,200 licencees and over 3 million experiments.[5] (The number of inspectors has recently risen to around 21 because, from 1987, animal breeders will also be registered.) The second line of defence is that without vivisection most of us would die, but we will come to that in a moment.

The familiar argument that most experiments involve nothing worse than a simple pin prick or a change in diet, conveniently ignores the possible *results* of these tests, or what is actually being administered. For instance one of the UK's best-known defenders of animal experiments, Birmingham's Owen Wade, wrote in 1983 that,

> 'Most animal experiments (3.7 million, or 80 per cent) are done without anaesthetics because feeding experiments, taking venous blood, or giving injections, do not require anaesthetics in animals any more than in man. In more extreme procedures anaesthetics are used and are as effective in animals as in man.'[6]

Tim Biscoe, Honorary Secretary of the Research Defence Society, presents an equally comforting view: '. . . we anaesthetize all vertebrates except under special and carefully controlled conditions for minor procedures.'[7] Perhaps the most revealing statements come from the pharmaceutical industry's Animals in Medicines Research Information Centre (AMRIC), set up to lobby MPs and journalists during recent debates on new legislation. AMRIC literature (leaflet code AMR/84/4) states: 'Anaesthetics must be used for potentially painful experiments.'

In the UK vivisection has been regulated for over a century by the Cruelty to Animals Act of 1876, which only covers experiments 'calculated to inflict pain', that is, potentially painful experiments. What AMRIC fails to say is that, if scientists wish to dispense with the anaesthetics, they need only obtain a certificate A from the Home Office. In 1986, 2,275,900 experiments 'calculated to inflict pain', or 73 per cent of the total, were carried out under certificate A.[5] Certificate B allows animals to recover from the anaesthetic should they be lucky enough to receive one in the first place. During 1986 583,262 experiments, or 18.7 per cent were performed under certificate B.[5] That leaves just 252,889 experiments, or about 8 per cent of the total, where anaesthetics were used throughout, with the animals not being allowed to recover consciousness.[5] AMRIC goes on to repeat Wade's assertion that,

> '80 per cent of experiments do not require an anaesthetic because they are not likely to be painful.'

All experiments carried out under the Cruelty to Animals Act are likely to be painful to some degree since the legislation only covered experiments 'calculated to inflict pain.' But industry seems to have forgotten the infamous LD50 test in which animals are deliberately poisoned to death and where even the Home Secretary's Advisory Committee acknowledged that 'LD50s must cause appreciable pain to a proportion of the animals subjected to them.'[8] In other more prolonged toxicity tests the highest dose level is deliberately chosen to *induce toxicity*[9] so that doctors supposedly know where to expect adverse effects during clinical trials with volunteers. Also, animal models of human disease rarely use anaesthetics yet often cause suffering as Dr Howell of Derbyshire's Royal Infirmary explained in the *British Medical Journal*:

> 'Unfortunately, much experimental work is performed on animals in medical research which must cause the animals prolonged suffering. This is true of most chronic experiments, such as the production of experimental models of human disease. Insofar as they are successful the animal must suffer, just as the patient suffers from the disease. Does a rat not suffer from a headache with an experimental brain tumour? The human counterparts of experimental alpha rigidity in dogs and cats are all excruciatingly painful, and the muscle spasms may fracture the neck of the femur.'[10]

Home Office statistics reveal that the vast majority of psychological experiments involving the deliberate infliction of stress or the use of 'aversive stimuli' such as electric shocks, are carried out without anaesthetics, as are a large proportion of burning and scalding tests.[5] Sadly there is no shortage of published examples.

A more subtle device, invented by the Home Office but used by experimenters and government ministers alike, is the pain condition – supposedly designed to prevent excessive suffering. This states that animals '. . . suffering severe pain which is likely to endure . . . shall forthwith be painlessly killed.'[11] But no one has ever defined 'severe' or 'likely to endure', and both are left to the scientist to define. Even the Home Office's own Littlewood Committee reported in 1965 that, 'It is not as a rule possible to assess degrees of real pain in animals' and '. . . it is often much more difficult to detect physical signs of pain in an animal than in a human patient.'[12] And the former Chief Inspector at the Home Office, Dr J. D. Rankin, admitted that,

> 'There is no way in which we can measure severity. There is no way in which we can measure endurance . . . That does not mean to say that everybody has the same view about what is severe and what is enduring . . . But each licensee must know himself what "severe" is and what "enduring" is. It is, however, and can only be, a subjective assessment.'[13]

As it is impossible to measure levels of pain in animals, it follows that what may be considered severe by one person may be dismissed as mild or moderate by another. For instance when former MP Frank Hooley questioned the electric shock-induced fighting experiments at the University of Bradford, he was told by Lord Jellicoe,[14] chairman of the Medical Research Council that funded the research, that the shocks were 'mild'. But the scientists themselves describe how the animals '. . . reacted to aversive stimulation with vigorous motor and vocal responses',[15] that is, they cried out and made desperate attempts to escape, which is hardly what you would expect from mild electric shocks.

The pain condition then, like similarly worded 'safeguards' attached to the latest British legislation, is utterly meaningless to the animals but does allow the government and experimenters to pretend there are strict controls against excessive suffering. When the writer, together with former NAVS General Secretary Brian Gunn, visited Home Office officials to discuss new legislation, we were told that LD50 tests did not breach the pain condition. The suffering, they assured us, was not severe, but substantial!

In any case, what faith can we have in scientists who must inevitably become so hardened to suffering and death that they are insensitive to pain in their animal victims? Dr Howell explains how Canadian neurologists who chose to spend a year of their training experimenting on animals, had so hardened themselves to animal suffering that they were no longer capable of recognizing suffering in their patients for quite a while after returning to clinical work.[10]

Even when anaesthetics are given, for instance in conjunction with muscle relaxants, how do we know the animal is completely free from pain? Muscle relaxing drugs leave animals physically helpless but fully conscious so they have no way of showing their distress. And, conveniently, there is no way they can complain after the experiment should the scientist be careless and give insufficient anaesthetic. Recently a High Court judge awarded £13,775 to a woman who, during an operation for caesarian section, could feel every cut of the knife because she had not been given enough anaesthetic.[16] She could not indicate her distress or even scream because a muscle relaxant had also been administered. In awarding damages against the Wigan Health Authority, the judge described the woman's ordeal as '. . . the most horrific experience I have ever heard.'[16]

In vision experiments carried out by Colin Blakemore and his colleagues, cats were paralysed with a muscle relaxant and given a mixture of nitrous oxide and oxygen as anaesthetic.[17] But other scientists have argued that nitrous oxide is not itself a general anaesthetic in cats and its use in conjunction with paralysing agents is based on a claim by Blakemore himself that it adequately maintains anaesthesia first induced with barbiturates. Other scientists argue that this claim is mistaken so the cats, although paralysed during the surgical operations, may not be adequately anaesthetized.[18]

As more and more photographic and documentary evidence becomes available, often only as a result of daring undercover raids, the claim that animals are well treated starts to wear thin. On August 26, 1984 members of the South Eastern Animal Liberation League entered the sacred research laboratories of the Royal College of Surgeons at Buckston Browne Farm in Kent and seized documents and photographs that were to form the basis of subsequent legal proceedings by the BUAV under the Protection of Animals Act 1911. The College's own incident reports revealed that caged monkeys used in dental research had been found dead, trapped by their arms, whilst a ten-year old macaque called Mone, used for breeding purposes, had been found in the intense heat of June 22nd, collapsed on the floor of her tiny 3 foot by 2 foot 6 inch by 2 foot 6 inch cell, severely dehydrated.[19] Press coverage was enormous as an incredulous public began to realize what really happens behind the closed, locked doors of UK laboratories. Even so, no case could be made that animals were suffering *during* actual experiments because, even if they were, this would have been sanctioned by the Cruelty to Animals Act. An Early Day Motion in the House of Commons called on the Home Secretary to make a formal statement about the role of the Home Office inspectors who had called at the Royal College only four days before Mone had collapsed of dehydration and on the very day that Rage, another female macaque, had actually died of dehydration.

> '*The Royal College of Surgeons and Home Office Inspectors*
> That this House notes that a monkey named Mone was discovered collapsed and severely dehydrated on 22nd June 1984 at the Royal College of Surgeons Research Establishment, Downe, Kent, and that on 18th June 1984 a Home Office inspector called at the centre and that on the same day a monkey, Rage, was discovered by staff severely ill and died soon afterwards from dehydration; is aware that over a six-year period a further 52 monkeys have been reported trapped in their cages by staff and a number have died as a result; and, noting that the Home Secretary is responsible for conditions in this and other laboratories where experiments on animals take place, calls on the Home Secretary to make a statement as soon as possible in the House as to his future intentions.'   (Early Day Motion No. 494)

On February 18, 1985, the Royal College of Surgeons of England was fined £250 for causing unnecessary suffering to a ten-year-old female macaque monkey used for breeding purposes. However, the ruling was ultimately overturned on what some argue was a point of law[21]: after an appeal by the RCS against the original conviction was dismissed, two High Court judges eventually quashed the conviction because the College had not been given the chance to answer new facts raised by the appeal court concerning the *reason* for the monkey's dehydration.

Nevertheless, further revelations soon followed giving a brief

INCIDENT REPORT

This form is to be used for all significant incidents involving animals at Downe – for example, escapes, injuries or disease. It should be filled in by the relevant member of staff and either the original or a photocopy forwarded to:- J. E. Cooper, Veterinary Conservator, Royal College of Surgeons of England, 35–43 Lincoln's Inn Fields, London WC2A 3PN (Tel: 01-405 3474).

Continuation Forms are available in both Units and should be used to monitor an animal's progress or other developments. The Continuation Form should remain at Downe until completed.

Date of incident: *17/5/84*   Time: *8.30*

Species of   Name or
animal: *Macaque*   number: *DRUDE* Sex: ♀

Status: ~~Stock~~/~~Breeding~~/   Age: -
Experimental

Nature of experiment: -   Responsible scientist: *D. B.*

Room number: *201*   Cage number: *Commune 5*

Type of incident: ~~Disease~~/Injury/~~Other~~

Details of incident: *Found hanging from top of cage by left arm (trapped) – Dead.*

Identifiable cause(s): *Unable to reach water/shock?*

Action taken: *Body removed quite easily from trapped position. Taken to theatre – samples taken (lung, liver, kidney) and general p.m. performed. Left wrist damaged.*

Notes: *Animal becoming trapped in cages is now fairly common especially communes. If and when new cages are brought areas where these incidents can arise should be avoided.*

INCIDENT REPORT

Date of incident: *22/6/84*   Time: *8.30*

Species of   Name or
animal: *Macaque*   number: *MONE*   Sex: ♀

Status: ~~Stock~~/Breeding/   Age: *10*
~~Experimental~~

Nature of experiment: -   Responsible scientist: -

Room number: *169*   Cage number: -

Type of incident: ~~Disease/Injury/~~ Other

Details of incident: *Animal found severely dehydrated.*

Identifiable cause(s): *Ventilation system inadequate. Each year during summer months temperatures in animal areas soar, this year they have regularly been between 85–92°F, they should be 68–72°F.*

Action taken: *Given 60 ml fluids sub/cut. and 100 mls by i.v. drip – Placed in heated cage. After recovery given complan and H_2O 60 mls and 60 mls.*

Further action needed: *Overhaul ventilation system. Place intake and extract vents in* correct *positions i.e. at high and low level. At present they are* opposite *each other which means there is no circulation whatever!*

These two pages are reset from photocopies of Royal College of Surgeons incident reports.

insight into the normally secret world of animal supply. Documents revealed that during the 1970s, the RCS received beagles from the Chemical Defence Establishment at Porton Down and primates from the Sussex-based Shamrock Farms. More recently, many dogs used in their research have been supplied by Hampshire vet David Walker through his APT Consultancy. They included alsatians, spaniels, collies, various mongrels, greyhounds and an Old English sheepdog.[20]

Mr Walker was also Director of Research at Wickham Laboratories, a contract house run by another vet William Cartmell, that carries out tests for the pharmaceutical, cosmetics, pesticides and chemical industries. Unpublished Wickham documents obtained in yet another raid by animal activists, clearly highlight the misleading nature of statements by the then Parliamentary Under-Secretary of State at the Home Office, David Mellor.

With new legislation before parliament to update the Cruelty to Animals Act of 1876, David Mellor was under considerable pressure to ban the Draize test. Instead he not only dismissed claims that the method was unreliable and misleading but gave the impression that it was relatively harmless too. In a letter to Sir Russell Johnston MP, dated November 27, 1985, David Mellor states that the test is stopped '. . . at a point at which you would be aware that something was irritating your eye but a long way from the point at which you could say that serious damage had been inflicted.' He goes on to say that suspect chemicals, which presumably means all substances so far untested, are '. . . usually instilled first in a weak solution to minimize the possibility of a serious reaction and damage to the eye.'

The Home Office must have known that most Draize tests are not published because they are carried out by industry or by contract research laboratories, so it would be hard to disprove his case with specific examples. The Wickham raid changed all that. In May 1982, the laboratory tested a series of Breox products that contain polyalkylene glycols used in brake fluids and some cosmetics. The products were supplied by British Petroleum and all were administered *undiluted* to groups of three rabbits, with the animals being observed for one week. One of the Breox products had a rapid effect.

> 'Within an hour of instillation the conjunctivae were conspicuously swollen and crimson red. The eyelids were half closed, inflamed and moistened by discharge . . . At the same

time the corneas were slightly but extensively opaque and the irides were congested but still reacted to light. On Day 1 the conjunctival scores decreased but the corneal and iridial reactions persisted.'[22]

The report notes that after seven days, all eyes had returned to normal. In March 1982, BP had again asked Wickham to carry out eye irritation tests, this time with a cutting oil used as a coolant and lubricant. Tests on *both* the undiluted and diluted product were requested although toxicity of the product had not apparently been investigated before.

'The rabbits were placed in neck-stock restrainers for dosing ... Ocular changes in animals with the *undiluted* test article were severe by comparison with those of other groups. Within an hour of dosing their conjunctivae were crimson red, obviously swollen and discharging. This conjunctivitis persisted and on Day 1 was accompanied, in all rabbits, by a corneal opacity which tended to obscure details of the iris.'[22] (emphasis added)

For 'humane reasons' the eyes were treated and returned to normal after four days. In August 1982 Wickham tested an *undiluted* mouthspray concentrate, which led to substantial irritation involving redness, swelling and discharge.[22] The irritation gradually declined but was still evident after seven days.

But it is not just Wickham: documents from Hazleton Laboratories in Harrogate reveal that undiluted paraquat solution was applied to rabbits' eyes during 1980, causing 'moderate' eye irritation.[23] Paraquat was already known to damage human eyes. The Taiwan company commissioning the tests, Sunlead Chemical Industry Co. Ltd., issued special handling precautions to safeguard Hazleton staff: 'Avoid contact with eyes ... Can be irritant to skin and eyes ... skin and eye contact, flush with copious quantities of water.'[23] If the chemical was known to be an eye irritant, why were the tests carried out at all? And why were they carried out with *undiluted* paraquat if, as David Mellor stated, suspect chemicals are '. . . usually instilled first in a weak solution to minimize the possibility of a serious reaction and damage to the eye'? Several of the Wickham laboratory examples too were carried out with *undiluted* products.

In earlier propaganda Mellor had stated that,

'The whole purpose of an eye test is to see whether any damage is done to the eye. Once it is detected, the test can stop. There is no need to make an animal suffer an irritant over a long period and propaganda that suggests that animals in this country linger for days with swollen bloodshot eyes is completely false.'[24]

Test chemicals can be classified into two groups: those that are designed to come into contact with the eye and should not therefore cause any eye irritation (contact lens solutions for example), and those that might be expected to cause damage from accidental exposure (pesticides, weedkillers, dandruff shampoos, detergents and many industrial chemicals) but that, nevertheless, are considered commercially important. In the latter case signs of damage to the rabbit's eye would be monitored over several days to see if they healed or deteriorated. After all, a stated purpose of the Draize test is not only to discover the type and severity of damage but also the time of onset, duration and possible resolution of any injuries or inflammation and whether the damaged tissues ever return to normal.[25] Consequently tests might be increasingly painful should the injuries deteriorate. An example already quoted in Chapter 8 is tear gas, where the corneal injuries actually deteriorated for seven days, after which they remained much the same for a further week.

At least the government does not deny that the Draize test is still carried out: it would be hard pressed to because Home Office statistics show that during 1986, 11,263 experiments were performed in which substances were applied to animals' eyes. Yet, during the recent Committee stage of the Animals (Scientific Procedures) Bill, Lord Halsbury, past President of the Research Defence Society, actually claimed the Draize test is no longer used: '. . . I must place on record that the Draize test is obsolete. It is no longer used . . . so why bring it up now?' (*Hansard*, 415, 12 December, 1985)

The Draize test is not confined to Wickham and Hazleton Laboratories. In 1984 university student Richard Beggs took a job as technical officer inside Herefordshire's Toxicol Laboratories at Ledbury. In the weeks ahead he would compile a horrifying report which showed that animals not only suffered in toxicity tests, but during their confinement as well.[26] In skin sensitization tests, which can result in sores on the animals' backs, five guinea pigs are kept in a plastic cage 2 feet by 1½ feet with a wire floor, no bedding or exercise, a pelleted diet and artificial light. He records how severe struggling can result in a broken neck when rabbits are being injected during pyrogen tests. On one occasion, he notes that, '. . . when the lower level of animals was pulled forward in the stocks, the whole assembly of stocks overbalanced and fell over, resulting in 15 broken necks.' In carcinogenicity tests, he notes that some rats were swollen to twice their normal size because of tumours. He also discovered some of the products tested. These included aerosols, dental cream, deodorants, shampoos, carpet cleaner, insect killer, foot cosmetics, sun lotions, antiperspirants, hair-care products and detergents.[26] Toxicol's clients include many household names and later that year the *Sunday People*, who in the mid 1970s had exposed ICI's smoking beagles, not only produced pictorial evidence of life inside a UK contract laboratory, they also named the top firms involved in animal testing – Colgate Palmolive, Johnson and Johnson, and Dunhill Toiletries.[27]

An even more dramatic case of a 'mole' at work came in September 1981 when Montgomery County police raided a medical research centre and seized documents, samples and 17 monkeys, which they took into protective custody.[28] When police entered the Institute for Behavioural Research in Silver Spring, they said that the monkeys were in such a state of physical and mental stress, they appeared to have bitten their fingers and arms. The scientist responsible, Edward Taub, a behavioural psychologist with 24 years experience in primate research, had cut the nerves to one arm of each animal in an attempt to duplicate the effects of a stroke. After the operation the monkeys were prevented from using their good limb to see how they managed with the other arm.

Four months before the police raid a young college student called Alex Pacheco took a job in Taub's laboratory. Once established, Pacheco, co-founder of the increasingly successful animal rights group PETA, photographed conditions in the monkey colony room and, when Taub was away on holiday, invited five experts in primate behaviour to visit the lab. Pacheco's photographs depict scenes described by one of the five observers, primatologist Dr Geza Teleki, as 'some kind of hell': dirty, rusting, immovable cages encrusted with mouldy excrement, dried blood on the walls, monkey food lying in filthy faecal trays, broken wires protruding into cages, the monkeys themselves with draining wounds and limbs with stumps where there had once been fingers. On Taub's desk Pacheco photographed a monkey's skull and hand serving as paper-weights. As a result of the photographic evidence and sworn affidavits by the five scientists, police obtained a seizure warrant from a Montgomery County Judge. Yet Taub denied conducting experimentation 'out of the ordinary' (*New York Times*, 13 September, 1981).

These are but some of the more dramatic actions that in recent years have helped focus public attention on the desperate plight of laboratory animals. The constant pretence that experiments cause little suffering only serves to undermine confidence in the scientific community and leads to suspicions about other claims for animal-based research. And here we come to the second line of defence, that of emotional blackmail, preying on the fear of disease to sustain public acceptance of the animal sacrifice.

The all-too-familiar argument is presented by the Executive Director of the Research Defence Society, James Noble:[29] the choice is simple, he says, which life is more important – that of a child or a soft-eyed spaniel? If pressed most people would choose their own child rather than someone they did not know but that would not justify experiments on strangers! Such hypothetical arguments are plainly absurd. What kind of medical system is it that demands a ritual sacrifice before healing the sick? Yet this is Claude Bernard's medieval doctrine, that '. . . we can save living beings from death only after sacrificing others' (see Chapter 5). In fact, the evidence presented here shows that the real choice is not between dogs and children, it is between good science and bad science; between methods that directly relate to humans and those that do not. By its very nature vivisection is bad science: it tells us about animals, usually under artificial conditions, and not about people.

Another device used by the RDS is to imagine the consequences had animal experiments been stopped at any given year in the past. For instance RDS literature states,

> 'If the antivivisectionists had succeeded in halting animal experiments in 1950, for example, hundreds of thousands of people around the world would have died from polio since then, and many more would have been disabled.'

For the research community this must be one of the most insulting arguments because it assumes that if one method is prohibited, scientists will simply give up, go home and put their feet up! As we have already seen in the case of the placenta and microsurgical practice, legal prohibition has entirely the opposite effect and provides the more enlightened with the incentive necessary to devise new techniques. In any case, when polio vaccines were first introduced in the 1950s, both here and in the United States, the disease was already in retreat whilst the most crucial breakthrough in preparing the vaccine came in 1949 when Enders and his colleagues showed how polio virus could be grown in human tissue culture.[30] For this they received the Nobel prize. So, if animal-based research had stopped in 1950, scientists would have been forced to prepare the vaccine from human cells, which we now know to be safer: Salk and Sabin both used monkey kidney tissue despite Enders' findings. Perhaps the RDS should amend its literature, with vivisection being stopped in 1930. Then Brodie's disastrous polio vaccine made from live monkeys would never have happened.[30]

The secrecy, the inflated claims for animal-based research and the frequent denial that tests cause anything more than mild discomfort, all point to powerful vested interests whose profits and livelihood depend on the survival of vivisection. Usually organizations defending vivisection keep a low profile except when new legislation on animal experiments is imminent – then they emerge from the woodwork!

In 1979 when American Congressmen put forward the Research Modernization Bill to divert 30-50 per cent of animal research money to the improvement of alternatives, one of the world's foremost laboratory animal breeding companies, Charles River, duly set up the Research Animal Alliance to lobby on its behalf. In 1980 a representative of Research Animal Alliance visited France and England '. . . to apprise certain groups of RAA's existence and purpose, to suggest liaison between these groups and RAA, and to assess the current anti-laboratory animal sentiments in Europe.' According to RAA's update for July 1980:

> 'While in England, RAA met with the working party on Laboratory Animal Regulations of the Associated British Pharmaceutical Industries (which is similar to the Pharmaceutical Manufacturers Associated in the United States). This group began five years ago in response to growing antilaboratory animal sentiments expressed in the United Kingdom and was formed to protect the interests of the pharmaceutical companies in this regard. RAA also met with

two representatives of the Research Defence Society (RDS), which has been in existence for many years and is dedicated to dealing with issues affecting the laboratory animal community. RAA found the Research Defence Society most receptive to the idea of working with RAA and exchanging information regarding regulatory and antivivisection activities in our respective countries . . .'

Since 1983 when the British government first published its proposals to update the Cruelty to Animals Act 1876, the drug industry has been particularly active. The Association of the British Pharmaceutical Industry, the drug industry's trade association, allocated a vastly increased budget for public relations for 1985. Top of the list of activities covered by the increased budget was public relations on animal experimentation, costing an estimated £125,000.[31] During a nation-wide referendum campaign to ban vivisection in Switzerland, the resident companies – Ciba Geigy, Hoffman-La Roche and Sandoz – were reported to have spent an estimated $9 million on propaganda.[32] Not surprisingly the vote was lost and the experiments continue.

Apart from the obvious commercial interests, many scientists base their entire careers on animal experiments. This is particularly true in physiology and experimental psychology. In medical research animal models of human disease become an end in themselves with researchers often completely divorced from the clinical situation. The former Director of Manchester's Trauma Unit, Professor Stoner, refers to the enormous number of physiologists who have studied a clinically irrelevant animal model of shock, arguing that they '. . . have no contact with 'real-life' medicine.'[33]

Another example is in blood pressure research. Clinical scientist Professor Colin Dollery explains that '. . . even for hypertension [high blood pressure] there really is no good animal model. There are plenty of animal models of hypertension but there is no really good model of human essential hypertension.'[34] Yet, since 1973 there have been *five* international conferences specifically on rat models of high blood pressure. The last one, held in Japan during 1985, contained 142 papers all on experiments with rats.[35] Despite all their efforts the important advances in treatment have come from clinical work: the blood pressure lowering effects of the main drugs used for treatment were first discovered in human patients[30] whilst clinical trials have also reported the beneficial effects of a vegetarian diet.[36] As long ago as 1904 Ambard and Beaujard observed that dietary salt restriction might ameliorate human hypertension.[37] This has since been confirmed by analysis of different populations and by clinical studies and, of course, the experiments have been repeated in animals.

With experimenters building their careers around vivisection any criticism is inevitably seen as a threat with researchers arguing strongly that their aim is really to help sick patients. For some this may be true but former RDS chairman David Smythe doubts if the desire to alleviate human suffering is often the incentive for research: 'The real motives', he writes, 'are a mixture in varying proportions of scientific curiosity, desire to explore new fields, desire for recognition and fame, career ambition, a wish to spend time deeply absorbed in something of special interest.'[38]

Career advancement is primarily dependant on publications: the more articles published, the better the chance of promotion. Writing in the *Lancet* Dr E. J. H. Moore notes that,

'The pressure on young doctors to publish and the availability of laboratory animals have made professional advancement the main reason for doing animal experiments.'[1]

'The procession of papers confirming the results of other studies', he writes, 'or adding some minute piece of useless information is depressing and dreary.'[1] Experiments on animals are ideally suited to the task of generating papers: for if one species proves inapplicable, another can be chosen, and another ... One doctor has described the laboratory rat as '... an organism which, when injected, produces a paper.'[39] Furthermore, differences between the species can be used to prove almost any theory. According to Dr Bross, Director of Biostatistics at New York's Rosewell Park Memorial Institute, 'The virtue of animal model systems to those in hot pursuit of the federal dollars is that they can be used to prove anything – no matter how foolish, or false, or dangerous this might be.'[40]

Perhaps the most disconcerting defender of vivisection is the government itself. Once licensed, the government is committed to defend the experiment otherwise it will lose face. It must also defend tests carried out in its own laboratories, such as Porton Down, as well as those it sponsors directly in contract laboratories or indirectly through the Medical, Science and

Engineering, and Agricultural Research Councils. The Department of Education and Science makes an annual grant to MRC and SERC whilst the Ministry of Agriculture, Fisheries and Food provides the cash for the ARC.

There are political and financial reasons too. Public concern over the thalidomide disaster forced government to seriously consider the way new drugs are tested and marketed. Following such a tragedy, failure to act would be politically damaging. Knowing that animal tests are intrinsically unsafe, the government could take a number of steps to prevent further disasters. It could introduce legislation limiting new drugs to those fulfilling a real medical need; it could urgently sponsor research into more reliable test systems (thalidomide's effects can indeed be seen in human tissue culture); and it could introduce a really effective system of patient monitoring once a drug has been marketed. The government did none of these things. Instead it introduced the 1968 Medicines Act, which simply resulted in more animal tests.

Lip-service was paid to post-marketing surveillance with the inept yellow card system, subsequently shown, as we have seen, to report only a tiny fraction of side-effects.[41] And the drug disasters continued: Eraldin, Opren, Flosint, Zomax ... Nevertheless government could claim that *something* was being done to protect consumers even if the tests were more in the nature of a public relations exercise: most people could not be expected to know that. It was also in the government's best financial interests to protect the pharmaceutical industry because of its large contribution to our balance of payments. So the drug industry was allowed to flood the market with unnecessary 'me-too' drugs that could only add to the burden of iatrogenic disease (side-effects from treatment).

The government was equally protective towards industry in 1983 when setting out its proposals to update the Cruelty to Animals Act 1876, the legislation controlling animal experiments in Great Britain.

> 'The United Kingdom has a large pharmaceutical industry which makes a large contribution to our balance of payments and employs 67,500 people. In devising new controls it is very important not to put industry at risk unnecessarily.'[42]

So it is not surprising that successive governments have traditionally defended even the most outrageous tests, such as the LD50, despite growing condemnation from the scientific community.

In 1979 the government's Advisory Committee on Animal Experiments, appointed by the Home Secretary, published its own report on the LD50 and concluded that the test should continue.[8] In response to a suggestion that the test be abolished or phased out, the report stated:

> 'Having regard to the worldwide use of LD50s, this is hardly a realistic proposal. But even if it were we would not support it.'[8]

Two years later, as criticism mounted, Britain's Department of Health and Social Security (DHSS) issued a further statement defending the LD50.[43] There was no attempt to rebut the scientific arguments, the DHSS simply stating that, '. . . any attempt to abandon the LD50 in view of its generally accepted use . . . is to swim against the tide and would not serve any useful purpose.'[43] But then the government really gave the game away in rejecting yet another call for the test to be abolished:

> 'LD50 results are required by drug regulating authorities in countries to which we export, contributing to a net favourable balance of trade in pharmaceuticals of over £500 million.'[44]

By 1984 intense pressure had forced the government to drop all mention of the LD50 from the Medicines Act test guidelines. Nevertheless companies are still asked to determine a new drug's lethal dose, thereby requiring at least an approximate LD50, albeit with fewer animals.

Since it is in the government's best interests to defend certain animal experiments, the Home Secretary must ensure his Advisory Committee maintains the status quo. In fact, the majority of members sitting on the Committee are either licensed to perform animal experiments or are very sympathetic to the method. Even the tiny contingent of animal welfare representatives has been carefully chosen by the Home Secretary. Two of the 'welfare' representatives represent organizations* that supported the government's own Bill to update the Cruelty to Animals Act of 1876, even though it fails to prohibit a single area of animal experimentation. In its battle for public and parliamentary

---

* FRAME (Fund for the Replacement of Animals in Medical Experiments) and CRAE (Committee for the Reform of Animal Experimentation).

approval, the government could therefore claim the support of 'moderate' animal welfare groups and give the impression that all is now well for laboratory animals, at the same time discrediting the UK's leading antivivisection societies that strongly opposed the Bill.

As the demand for animal tests increased – not only for drugs but pesticides, weedkillers, cosmetics and household and industrial products – business has boomed for the contract laboratories, the animal breeders and the cage and equipment suppliers. One of the UK's largest contract research organizations is Huntingdon Research Centre. For the year ending September 30, 1984 revenues exceeded £17 million, an increase of 12.6 per cent over the previous year whilst net income was £2,308,000, 61.7 per cent higher than in 1983. Overall HRC received contracts from 452 different clients during 1984 with most of the work coming from overseas:[45]

	1982 (per cent revenues)	1984 (per cent revenues)
UK	25	21
Europe	58	52
Japan	8	17
USA	6	8
others	3	2

Apart from routine toxicity tests, HRC has been responsible for some particularly distressing experiments in which rhesus monkeys have been poisoned with arsenic,[46] and tetra ethyl lead, the petrol additive.[47]

Another of the UK's major contract stations is Hazleton Laboratories in Harrogate. The company carries out tests for the drug, agrochemical, tobacco, cosmetics and household products industries but is only one part of a world-wide network centred in America. Apart from the UK, Hazleton Laboratories Corporation has test laboratories in Virginia, Wisconsin and West Germany. It claims to be the world's largest independent biological and chemical research organization and employs over 1,000 scientific and technical staff for this part of its business. In 1982, biological and chemical research contracts accounted for

63 per cent of its revenue, overall estimated at $53 million.[48]

Aside from routine toxicity tests, Hazleton's Harrogate laboratories claim to have special expertise: staff '... regularly undertake non-standard tests in the inhalation field and have particular expertise with potentially explosive gases'[48]! Apart from the usual laboratory animals, Hazleton has a wide variety of birds and fish to test industrial and agricultural chemicals.

CONVULSION METER
MODEL: "CONVULS — 1"

Columbus Instruments Animal Convulsion Meter model "Convuls-1" is intended for objective, quantitative measurement of lab animal convulsions (rats, mice, gerbils, guinea pigs, etc.)

Measurement requires human participation, as it is impossible using economically reasonable means, to differentiate automatically between animal locomotor movements and convulsions

Animal is placed into a small transparent cage of which bottom constitutes precision Columbus instrument force platform. Convulstion Meter is activated by operator (pushbutton) each time animal starts convulsing

Electrical signals proportional to amplitude of animal convulsions are amplified and integrated over time. Digital display shows figure proportional to product of convusion force (in grams) multiplied by time. If results of measurements are divided by animal's weight, then such "weight compensated" convulsion measurements can be compared between animals of defferent size and expressed in grams of convulstion force per gram of animal's body weight. Printing Counter PC-800 or "Apple/IBM—PC Counter" interface to Apple or IBM—PC computer can be used to document results versus time.

ORDERING INFORMATION:

Catalog #167 — Convulsion Meter model "Convuls — 1-1" for single animal, complete including force platform and control unit. When ordering please specify size of animals.

Catalog #168 — Single channel printer, prints automatically in pre-set time intervals totalized convulsion force

Catalog #169 — Four animals convulsion meter model "Convuls — 1-4"

Catalog #170 — Printer for "Convuls — 1-4"

From Columbus Instruments catalogue

These include Japanese and Bobwhite Quail, Light Sussex Hens, Ring Necked Pheasants and Starlings. Rainbow Trout, Mirror Carp and Fathead minnows are amongst the species of fish available.

The pharmaceutical industry is Hazleton's biggest profit earner (51 per cent in 1983) followed by the agrochemical industry with 31 per cent and the tobacco industry with 11 per cent. The company also does a small amount of work on commission from the government. According to an internal memo, Hazleton has seen an increase in revenues from testing cosmetics and toiletries because of a lowering of its prices compared with other contract laboratories:

'The marked increase in cosmetics/toiletries revenues in F83 is a reflection of our policy of reducing acute study prices to Safepharm and Toxicol clients.'[48]

To supply the needs of contract houses, the drug, cosmetics, pesticides and chemical industries, as well as university and government laboratories, a vast network of laboratory-animal breeders has emerged. The 1984 catalogue of Interfauna, which incorporates three of Europe's best-known breeders, including Britain's Hacking and Churchill, and Shamrock Farms, lists the price of commonly used species:[49]

Wistar rats (250 grams)	£2.88
Mice (NIH strain) up to 5 weeks old	£1.40
Guinea pigs (300-350 grams)	up to £6.30
Rabbits (2-2½kg)	£15.00
Beagle dogs, 3-4 months old	£193
8-9 months old	£286
Cats	from £105

Stocks of rats, mice and guinea pigs are all 'established by hysterectomy derivation' to ensure sterility. Furthermore animals can be surgically prepared according to the customer's special requirements: castrated rats or animals with their adrenal, pituitary or thyroid glands removed are all available. Although most animals are specially bred, others are captured in the wild. In the case of primates, most die during capture and transportation.[50] Shamrock Farms, the Sussex-based company dealing in primates for research ('we are wholly ethical', says their advertisement), recently imported 10,000 wild-trapped cynomolgus monkeys over a four year period from Malaysia, Indonesia and the Phillipines.[51] They were shipped by air to the UK in batches of 50 via transit camps in the country of origin. The company reports how epidemic enteritis in the 10,000 newly imported animals was common, often posing a threat to life.

'Enteric disease is clinically and economically important in newly-imported simians and is induced by a combination of stress, malnutrition, parasite burden and infection with pathogenic bacteria.'[51]

It is difficult to imagine a more stressful experience for these animals than being captured and transported from their natural habitats and separated from their social groups. On arrival at the quarantine area the animals are placed singly in galvanized steel cages and each batch isolated for at least eight weeks. During the four-year period covered by the report, 192 batches of the cynomolgus monkeys were imported and all included animals with enteritis of '. . . varying severity, accounting for 33 per cent of deaths.'[51] (A further 46 per cent of deaths during this period were due to pneumonia.) Signs included sudden death or severe dysentery with profuse diarrhoea. Animals recovering from the ordeal of capture, transportation and isolation must then face whatever the laboratory has in store for them.

Animals have also been supplied from less orthodox sources. Documents obtained from the former London-based Animal Suppliers Ltd., a company supplying laboratories and the pet trade, suggest that zoos have been involved, albeit indirectly, in providing animals for vivisection. The evidence was obtained by the Central Animal Liberation League and shows that Ravenstone Zoo supplied the company with rhesus monkeys and baboons; Chessington Zoo supplied them with rhesus and African green monkeys, bonnet and stump-tailed macaques and capuchins; and Windsor Safari Park provided baboons.

Ex-racing greyhounds have proved popular in research,[52] whilst even the Wildfowl Trust at Slimbridge has supplied birds for experiments at Birmingham University.[74] Meanwhile the extent to which domestic animals are stolen for research is unknown.

With so much at stake it is little wonder that medical arguments against vivisection are ignored. Even the choice of species is largely dependent on economics. The *Textbook of*

Adverse Drug Reactions explains how the processes of absorption, distribution, metabolism and excretion vary 'extensively' from species to species and notes that,

> 'Difficulties of interpretation are compounded because the species routinely used in toxicological studies are chosen not on consideration of their phylogenetic relationship to man but on practical grounds of cost, breeding rate, litter size, ease of handling, resistance to intercurrent infection, and laboratory tradition.'[53]

So it is no coincidence that the animals most commonly used in research, including mice, rats, rabbits and guinea pigs, are also amongst the cheapest.

Ironically, mounting costs caused by an increased demand for tests can also prove an important incentive for reform. Traditional animal cancer tests to determine the safety of new drugs and other chemicals, are not only time-consuming but hugely expensive, which led to the rapid development of numerous *in vitro* techniques such as the Ames test. By comparison, other animal tests are relatively cheap so pressure to eliminate them comes mainly from the animal rights movement: in 1982 skin and eye irritancy tests cost as little as £45 whilst a routine rat oral LD50 only cost around £305.[54] Sometimes 'alternatives' face a financial *disincentive*. For instance, an effective vaccine against leprosy can be made from human tissue but this is considered too expensive so, instead, the vaccine is produced from deliberately infected armadillos.[55] Another obstacle is the inflexible attitude of government regulatory bodies who demand animal experiments for new drugs. They are very loath to eliminate any test, however meaningless it may be, just in case there is a disaster: the blame would then fall on them. Dr Dunne of the World Health Organisation explains why government agencies refuse to scrap tests that have proved ineffective:

> 'The agency sees itself as building a wall of bricks with every test a brick – as a dam to stop any seepage of foul water past it. If the agency sees a hole in the dam it doesn't replace one brick by another – it adds another layer of bricks.'[56]

So, any newly developed test is likely to be *added* to the check-list rather than replace an existing procedure. Inevitably this deters a cost-conscious industry from introducing more reliable toxicity tests, such as those based on human tissue.

The initial cost in switching to human-based medical research might also be seen as a deterrent. In 1985 the author organized an informal meeting of scientific experts to see what could be done to encourage the wider use of human tissue in medical research, testing and training. Participants agreed that this was desirable both for humane and scientific reasons. One suggestion was for laboratories to establish human tissue banks up and down the country so material would be more readily available, the project being financed by the Medical Research Council. One of the participants had already set up a bank of human liver tissue at the Royal Postgraduate Medical School where samples are stored at −80°C until they are needed for research. But such initiatives cost money, even if they do eventually pay for themselves by reducing the number of animal experiments. In fact, when established, human-based research is likely to be far more economical. An expert in drug design writes:

'No laboratory animal will ever be a completely satisfactory substitute for the human system and the time will come when we shall stop wasting the enormously valuable enzymes and organelles of the dead and instead put these to use to understand the living human being better.'[57]

The introduction of human tissue banks for medical research would make an important contribution and could be reinforced by the MRC adopting a formal policy in which priority is given to grant applicants using human cells and tissues rather than animals. In the competitive search for grants this could bring dramatic changes and be part of a 'Campaign for Real Science'.

In recent years the 'alternatives issue' has assumed a prominent position but not just in defence of animals: it has also been cleverly used to disarm critics. Companies have been able to divert criticism about their use of animals by claiming to subscribe to organizations promoting alternatives. This may be true although there is no evidence that firms are setting up human tissue banks. Nevertheless, for public relations purposes, the message is clear: we are interested in alternatives so get off our backs! But if the industry was really concerned about reducing animal experiments, it would stop flooding the market with products for which there is no real medical need. With the great majority of new drugs adding little or nothing to those already available, we estimate that over 40 per cent of all experiments on animals could be eliminated by adopting the essential drugs policy described earlier.

Much the same is true of other industries: by focussing too much attention on alternative methods of research, more fundamental areas are often overlooked. For instance is there an urgent social or medical need for the product in the first place, and why is the research being done? Testing cosmetics and toiletries on animals is unnecessary because new products can rely on the many existing ingredients already shown to be safe through years of human experience. In fact, many companies marketing products as cruelty-free regard this as an important selling point.[58] New cosmetic formulations based on these well-established ingredients can still be developed, using patch testing with human volunteers to detect any skin irritancy.

Animal experiments to select, develop or study the use, hazards or safety of various substances (1986)[5]	
Cosmetics and toiletries	15,652
Food additives	8,988
Pesticides	55,301
Herbicides	23,565
Industrial chemicals	72,150
Household products	9,309

Again, why is it necessary to develop more synthetic food additives? Of the 3,500 food additives currently in use 3,000 are merely food flavourings[59] yet their safety cannot be guaranteed by animal tests. Only years of human experience will reveal the true dangers and by then it will be too late. For products that are not essential, is it really worth the risk?

Pesticides too are big business: the 29,500 tons used in the UK during 1984 were valued at £345.9 million.[60] Pesticides may be valuable to farmers in the short term but their ultimate effect is actually to *create* pests! This is because insects quickly develop resistance, requiring ever more powerful insecticides to destroy them. According to the National Academy of Sciences, pest damage in the United States increased from an estimated

$8.4 billion (31.4 per cent of the total crop) between 1940 and 1950, to an estimated $ 9.9 billion (33.6 per cent of the total crop) between 1951 and 1961. Figures from America's Environmental Protection Agency reveal that US farmers today use 12 times the amount of pesticides they did 30 years ago, yet the proportion of crops lost before harvest has almost doubled![61] The escalating use of pesticides has medical consequences as well, with disease-carrying organisms such as the mosquito also developing resistance. In fact, the World Health Organisation has warned that resistance is probably the biggest single obstacle in the fight against vector-borne diseases and is mainly responsible for preventing the successful eradication of malaria: '. . . evidence has also accumulated to show conclusively that resistance in many vectors has been caused as a side-effect of agricultural pesticide use.'[62] It is thought that over-use of pesticides on crops led to the re-emergence of malaria in India and Central America. The *Indian Journal of Public Health* actually refers to the uncontrolled use of pesticides for agricultural purposes as '. . . the most serious threat to public health.'[63]

By their very nature pesticides are poisonous. According to the London Food Commission, at least 49 pesticides have been linked with cancer; at least 31 with birth defects; and at least 61 with genetic defects.[60] They may also be connected with asthma, food intolerance and skin allergies. It is thought that 375,000 people in the Third World are poisoned – 10,000 fatally – by pesticides every year.[64] To make matters worse pesticides like DDT are very persistent. A survey of fruit and vegetables carried out in 1983 found that a third of the samples contained detectable residues of pesticides.[60] Residues have also been found in human breast milk.[65] So, in this case, the answer to animal testing is not to promote *in vitro* alternatives but to end our dependence on an ultimately ineffective and harmful system of pest control.

Another area where the alternative is simply not to do the experiment is the use of animals in agricultural research. The Agricultural Research Council is a major source of government funds for such research and one of the Council's principal animal users is the Institute of Animal Physiology at Babraham in Cambridge. The Institute's main function is to '. . . extend knowledge of the basic physiology and biochemistry of farm livestock as a foundation for improved animal production.'[66]

Intensive farming practices are already an enormous source of animal suffering yet reports from the Institute show that many of the experiments are designed to make these methods more productive.[66] The Director has explained that an '. . . important part of the programme is concerned with the motivational systems determining the needs of livestock for food, water, warmth, light, sex, etc.'[66] For instance, experiments with sheep and calves have been carried out to understand their illumination preferences. Such knowledge '. . . is useful in the selection of lighting levels for animals kept in intensive husbandry units.'[66] Lambs have also been subjected to maternal deprivation in experiments designed to study the mother-infant relationship.[67]

Because the animals are kept in such overcrowded conditions, intensive production methods cause tremendous stress and so-called 'vices' result. In the case of battery hens this involves feather pecking and cannibalism, so many producers routinely debeak birds as a preventive measure. Babraham scientists have tried to reproduce the vices experienced by factory farmed pigs by dosing the animals with drugs. Then they attempted to induce the same behavioural problems in guinea pigs and rats.[66] So an appalling system of food production leads to more suffering in the laboratory. There is no escape for the animals.

Furthermore we are told by the Research Defence Society that vivisection also benefits animals, yet veterinary medicines for cats and dogs represent only a minute fraction of the 'animal health products' administered to factory farmed animals. In his book *Assault and Battery*, Mark Gold describes how farming magazines and trade journals are packed with advertisements extolling the virtues of a multitude of substances as new drugs proliferate onto the market:

> 'Anti-stress sedatives; energy boosters to encourage animals to eat more; hormones to encourage speedy fattening, to regulate when animals give birth or to encourage conception rates, and antibiotics to fight mounting problems with disease: to the drug industry, factory farming is good news.'[68]

Without the drug industry the stressed and overcrowded conditions on factory farms would surely lead to an epidemic of infectious disease. In the United States more than 40 per cent of the antibiotics and other antibacterial products produced annually are used as animal feed additives and for other non-human purposes. Nearly 100 per cent of poultry, 90 per cent of pigs and veal calves and 60 per cent of cattle receive antibacterial drugs in their feed.[69] Once again animal experiments are used to sustain another area of animal abuse.

With so much at stake and attitudes so entrenched, moral, scientific and medical arguments *in themselves* are unlikely to be successful in ending animal experiments. But if the vested interests are immune to logic, they must eventually succumb to public pressure. Already the growing hostility to vivisection has sent shock waves through the scientific community. Reactions have been predictably hostile or defensive. In a letter to the *Times*, a former Director of Wellcome Research Laboratories actually wrote: 'Discount man's use of animals over the ages and we would still be living in a barbarous, unhealthy and probably cannibalistic society.'[70] William Cartmell of Wickham Research Laboratories, a contract house in Hampshire, stated in the *Mail on Sunday* that, '. . . we vets view the animal welfare/conservationist lobby as motivated by sinister political, anti-British organizations.'[71] But others have accepted the need for reform: this is no longer an issue that can be swept under the carpet. In 1986 the *Lancet* noted that the '. . . political momentum of animal welfare issues is growing year by year' whilst '. . . for increasing numbers of young voters, animal rights belongs not on the fringes of the political debate, but right at the heart of it.'[72]

Some even acknowledge the positive impact on medical research. Introducing a US conference organized by the Tissue Culture Association, Dr John Petricciani of America's National Institutes of Health argued that '. . . science and scientists need to look beyond the confines of animal experimentation.'[73]

> '. . . regardless of one's personal views on animal rights issues, I am convinced that the movement has had, and will continue to have, positive effects within the scientific community simply from the point of view that it makes us look at alternate experimental approaches, and at least in some cases that has meant moving into *in vitro* systems with the surprising realization that *in vitro* systems sometimes offer advantages of increased economy, rapidity, and sensitivity when compared with animal models.'[73]

Like other social evils, such as slavery, child labour and the suppression of women's rights, vivisection is all about power: the power to capture and subject normally sociable animals to a life of solitary confinement; the power to poison, burn, maim and blind and, ultimately, the power to kill. Every hour of every day another 20,000 animals die in the world's laboratories. Unlike oppressed humans, animals cannot help themselves. They cannot lobby MPs, organize demonstrations and pickets or vote out governments . . . But we can.

One thing is certain, vivisection has thrived through a conspiracy of secrecy and deception. At least we can begin by changing that.

1 E. J. H. Moore, *Lancet*, 975, 26 April, 1986
2 Half-hour tape digest available from Scottish Antivivisection Society, 121 West Regent Street, Glasgow; see also reference 3
3 Report by J. Robins, *Animals' Defender*, 81-82, November/December, 1985, and M. Balls, *FRAME News*, no. 7 August/September, 1985
4 See Chapter 2
5 'Statistics of experiments on living animals,' 1986, Home Office (HMSO, 1987)
6 O. L. Wade, *BMJ*, 1599-1600, 21 May, 1983
7 T. Biscoe, *RDS Newsletter*, October, 1983
8 Report on the LD50 Test, Advisory Committee on the Administration of the Cruelty to Animals Act 1876 (Home Office, 1979)
9 Official Journal of the European Communities, 19 September, 1984
10 D. A. Howell, *BMJ*, 1894, 11 June, 1983
11 Reproduced in reference 5
12 Home Office Departmental Committee of Inquiry on Experiments on Animals 1965 (HMSO)
13 J. D. Rankin, British Association for the Advancement of Science Symposium, 26 January, 1982, proceedings published
14 Letter to Frank Hooley, MP, from Lord Jellicoe, 30 June, 1982
15 R. J. Rodgers and R. M. J. Deacon, *Physiology & Behaviour*, 183-187, volume 26, 1981
16 Reproduced in *Animals' Defender*, November/December, 1985
17 H. M. Eggers and C. Blakemore, *Science*, 264-267, 21 July, 1978; C. Blakemore and G. F. Cooper, *Nature*, 477-478, 31 October, 1970
18 R. Drewett and W. Kani in *Animals in Research*, D. Sperlinger (Ed.) (Wiley, 1981) and references therein
19 Report by J. Gould in *Animals' Defender*, 37-39, May/June, 1985
20 Photocopies of animal supply documents prepared by D. Walker of APT Consultancy; see also *Animals' Defender*, 110-112, November/December, 1984
21 *Liberator*, April/May, 1986 (BUAV Journal)
22 Examples taken from photocopies of Wickham documents
23 Based on photocopies of documents from Hazleton Laboratories and recorded in *Animals' Defender*, 34-36, March/April, 1984

24 D. Mellor, letter to MPs dated 17 December, 1985
25 D. W. Swanston in *Animals and Alternatives in Toxicity Testing*, M. Balls, R. J. Riddell and A. N. Worden (Eds) (Academic Press, 1983)
26 Recorded in *Animals' Defender*, 91-94, November/December, 1984
27 *Sunday People*, 21 October, 1984
28 *New York Times*, 13 September, 1981; see also *Animals' Defender*, 14-15, January/February, 1982; *Nature*, 579, 22 October, 1981; *New Scientist*, 672-674, 3 December, 1981
29 Interview with M. Cotton *(Peterborough) Evening Telegraph*, 15 February, 1985
30 See Chapter 6
31 *SCRIP*, 1, 15 October, 1984
32 M. Studer, *Wall Street Journal*, 25 November, 1985
33 H. B. Stoner, *Advances in Shock Research*, 1-9, volume 2, 1979
34 C. T. Dollery in *Risk-Benefit Analysis in Drug Research*, J. F. Cavalla (Ed.) (MTP Press, 1981)
35 *Journal of Hypertension (Supplement)*, Supplement 3, volume 4, 1986
36 B. M. Margetts, et al, *BMJ*, 1468-1471, 6 December, 1986
37 L. Ambard and E. Beaujard, *Archives of General Medicine*, 520-533, volume 1, 1904
38 D. H. Smyth, *Alternatives to Animal Experiments* (Solar Press and the RDS, 1978)
39 A. R. Mitchell, *Lord Dowding Fund Bulletin*, 29-33, no. 16, 1981
40 I. D. J. Bross, 'Animals in Cancer Research' *The Antivivisection Magazine*, page 6, November 1982
41 See Chapter 3
42 *Scientific Procedures on Living Animals*, May, 1983, Command 8883 (HMSO)
43 J. P. Griffin, *Archives of Toxicology*, 99-103, volume 49, 1981
44 *Hansard*, 368-369, 24 November, 1981
45 *SCRIP*, 12, no. 948, 1984
46 R. Heywood and R. J. Sortwell, *Toxicology Letters*, 137-144, volume 3, 1979
47 R. Heywood, et al, *Toxicology Letters*, 11-16, volume 2, 1978
48 Hazleton information based on 1982 Annual Report and photocopies of documents obtained by animal rights activists (*Animals' Defender*, January/February, 1984)
49 Interfauna price list, March 1984
50 R. Ryder, *Victims of Science* (National Antivivisection Society and Centaur Press, 1983)
51 G. W. Tribe and M. P. Fleming, *Laboratory Animals*,, 65-69, volume 17, 1983
52 *Animals' Defender*, September/October, 1983 and January/February, 1984 and references therein
53 D. M. Davies (Ed.), *Textbook of Adverse Drug Reactions* (Oxford University Press, 1977)
54 Safepharm (contract testing laboratory) price list, 1982
55 *Nature*, 527, 18 June, 1981
56 Recorded in T. Smith, *BMJ*, 1333-1335, 15 November, 1980
57 C. Hansch in *The Use of Alternatives in Drug Research*, A. N. Rowan and C. J. Stratmann (Eds) (Macmillan, 1979)

58 BUAV and Animal Aid produce lists of companies producing cruelty-free cosmetics
59 E. Millstone, *New Scientist*, 20, 18 October, 1984
60 *Lancet*, 633, 15 March, 1986
61 E. Goldsmith, *The Ecologist*, 94-97, no. 3, 1980
62 G. Chapin and R. Wassestrom, *Nature*, 181-185, 17 September, 1981
63 Reproduced in reference 62
64 Reported by C. Canfield in *New Scientist*, 15-17, 16 August, 1984
65 *New Scientist*, 16, 13 March, 1986
66 Evidence presented by R. Sharpe on behalf of NAVS to the House of Commons Select Committee on Agricultural Research Spending, reproduced in *Animals' Defender*, 62-63, July/August, 1983, and references therein
67 BBC2 Horizon documentary programme, 5 November, 1984
68 M. Gold, *Assault and Battery* (Pluto Press, 1983)
69 J. Mason and P. Singer, *Animal Factories* (Crown Publications, 1980)
70 J. Vane, *Times*, 10 September, 1984
71 W. Cartmell, *Mail on Sunday*, 12 December, 1982
72 D. McKie, *Lancet*, 513, 1 March, 1986
73 J. C. Petricciani, *Lord Dowding Fund Bulletin*, 26-29, no. 17, 1982
74 P. J. Butler and A. J. Weakes, *Journal of Applied Physiology*, 1405-1410, volume 53, 1982

Postscript

Vivisection is not dead. Not yet.

Over a hundred years ago Robert Koch introduced a set of rules for establishing whether a specific germ caused the disease under investigation. One of his postulates, as they were called, stated that the microbe should reproduce the same condition in laboratory animals. The method was soon found to be unreliable, even by Koch,[1] so he could not have expected his system to survive. Yet soon after doctors recognized the first cases of AIDS in 1981, scientists were busy trying to reproduce the disease in laboratory animals. In 1984 the *Lancet* reported attempts by American researchers '. . . to transmit a severe, lasting immunodeficiency or clinical disease to chimpanzees and monkeys.' Rhesus, cynomolgus, stumptail and capuchin monkeys, together with chimpanzees were all inoculated with tissue from AIDS patients but none developed AIDS up to a year after they were infected.[2] By this time scientists in France and the United States had isolated a new virus from the blood of AIDS patients thereby confirming suspicions that the disease is caused by a virus.

In 1986 the *Journal of Virology* reported further attempts to induce AIDS in chimpanzees.[3] On this occasion the virus was administered to eight chimpanzees reared in a colony at the Yerkes Regional Primate Centre at Emory University in Atlanta. Observations continued for up to 18 months, but once again there were no signs of AIDS.[3] Nevertheless chimpanzees are favourite animals in AIDS research because they are the only species to maintain the virus in their bodies when injected with material from human patients. But they do *not* develop the disease. Indeed, N. W. King of Harvard's New England Regional Primate Centre reports that no experimentally infected champanzee had died of AIDS after more than two years.

'Because of these limitations and the fact that the chimpanzee is endangered and in extremely short supply, this species will probably be of limited value to AIDS research. Attempts to infect other species of nonhuman primates including those of the genus *Macaca* with HTLV-III/LAV [the AIDS virus], have been uniformly unsuccessful.'[4]

Whilst scientists work hard to develop an animal model, it is once again clinical investigation of patients and test tube experiments that have shown how to tackle AIDS. Epidemiology revealed how the disease spread and how it could be prevented through changes in behaviour. Once the virus had been isolated from the blood of AIDS victims, clinical observation of patients and test tube studies showed how it worked to weaken the body's defences. Dr Richard Tedder, a consultant virologist at the Middlesex Hospital pointed out that,

'Infecting the chimpanzee has very little value. If you want to follow the progress of this disease there are sufficient humans already infected. And is what goes on in a chimp's body relevant to what goes on in the human body?'[7]

Understanding the way in which the virus worked led, in turn, to an *in vitro* system using human cells for testing new drugs against the disease.[5] Drugs active in the test tube have already been submitted for clinical trial.

So history repeats itself. In the 1880s Koch had failed to induce cholera in a wide variety of laboratory animals and eventually had to rely on clinical observation of patients and microscopic analysis to pin down the guilty microbe. His careful (human) observations showed how the disease spread and how it could be prevented.

Even so prevailing dogma still demands an animal model of AIDS. So the chimpanzees who normally live happily in social groups will be kept in solitary confinement for years to see if any clinical signs of the disease develop.[6] This is vivisection.

1 See Chapter 5
2 D. C. Gajdusek, et al, *Lancet*, 1415-16, 23 June, 1984
3 P. N. Fultz, et al, *Journal of Virology*, 116-124, volume 58, 1986
4 N. W. King, *Veterinary Pathology*, 345-353, volume 23, 1986
5 See Chapter 6
6 Video film of chimps used in AIDS research at the SEMA Corporation laboratories in Rockville, Maryland, shows how the animals are kept singly in small, solid metal isolation chambers. Video available from BUAV in the UK and PETA in the USA
7 *Daily Mail*, 28 August, 1985

Appendix

Recent Examples of Animal Experiments in the United States of America

1 Beagles Burned

Experiments were carried out in which dogs were deliberately burned to test the usefulness of an anti-infection vaccine when given after severe burn injuries. Forty beagle puppies were clipped of hair from the neck to the base of the tail. Then the researchers marked with indelible ink the parts of the puppies' bodies to be burned. The burns were inflicted over a third of the anaesthetized animals' body surface using kerosene-soaked gauze which was ignited, allowed to burn for 60 seconds, and then extinguished with a wet towel. For 18 hours following the burning, the animals received 'light anaesthesia.' Ninety days later, only 16 per cent of the control animals had survived compared with 48 per cent of those given the vaccine. All the dying animals showed evidence of burn wound sepsis. (J. D. Stinnett, et al, 'Improved Survival in Severely Burned Animals Using Intravenous Cornybacterium Parvum Vaccine Post Injury', *Surgery*, 237-242, volume 89, 1981. Experiments carried out at Shriners' Burn Institute, Cincinnati and the University of Cincinnati Medical Centre.)

2 LSD and Cats

Experiments were carried out to see how the behavioural effects of LSD in young cats compared with those previously reported for older animals. Apart from the reported effects of LSD in humans, the authors admit that the behavioural effects in animals '... have also received monumental attention, and literally thousands of studies have dealt with this issue'.

Nevertheless they argue that virtually no one has compared the reactions of young individuals with those of adults. Seventy-one kittens aged between 4 and 112 days received several injections of LSD and showed uncoordination, tremors, vomiting, hair standing on end, salivation, dilated pupils, headshakes, limbflicking and abortive grooming. The authors concluded that LSD '. . . produced a constellation of behavioural signs that has been previously described in detail for the adult cat.' (M. E. Trulson and G. A. Howell 'Ontogeny of the Behavioural Effects of Lysergic Acid Diethylamide in Cats', *Developmental Psychobiology*, 329-346, volume 17, 1984. Experiments carried out at University of Texas, Dallas.)

3 Mice Die From X-Rays

Experiments were performed to see if a mixture of two drugs enabled mice to survive longer after exposure to X-rays. The tests were part of research into drugs that make tumours more sensitive to X-rays whilst at the same time protecting normal cells. In order to compare the survival of animals receiving the drugs with those that did not, the authors describe a technique called the oral radiation death syndrome in which the head and neck of a mouse are irradiated. The radiation destroys the skin lining the mouth and throat, making it too sore to swallow. The animals generally die from 'nutritional insufficiency' 10 days after the dose of X-rays. If the mixture of drugs protect against damage caused by irradiation, a higher dose of X-rays is needed to kill the animals. The procedure is described by the authors as giving '. . . an excellent endpoint.' (P. Grigsby and Y. Maruyama, 'Modification of the Oral Radiation Death Syndrome with Combined WR-2721 and Misonidazole' *British Journal of Radiology*, 969-972, volume 54, 1981. Experiments carried out at University of Kentucky Medical School, Lexington.)

4 Dogs' Tracheas Ignited by Laser

Experiments were carried out to investigate the lung and tracheal damage caused by laser-ignited endotracheal tube fires. These tubes are increasingly used in clinical practice during laser surgery but have to be used with caution. Mongrel dogs were anaesthetized and a tube made from either polyvinyl chloride (PVC), Rusch red rubber or silicone, inserted into their

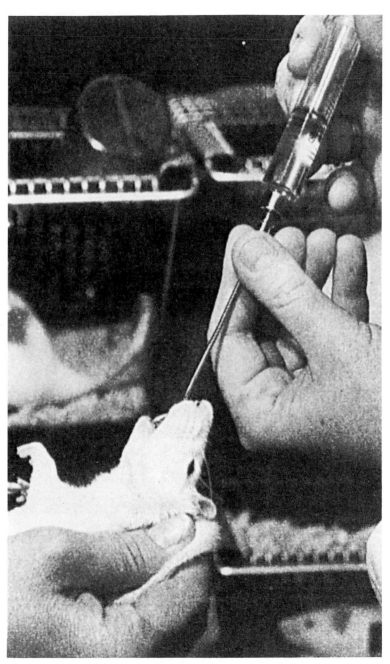

How the LD50 test is performed.

Left *A long-term toxicity study at Herefordshire's Toxicol Laboratories.* (© Richard Beggs)

Below *Destined for research. This cat was photographed whilst being held at Sheffield's Lodge Moor animal house.* (© Sue Merriken)

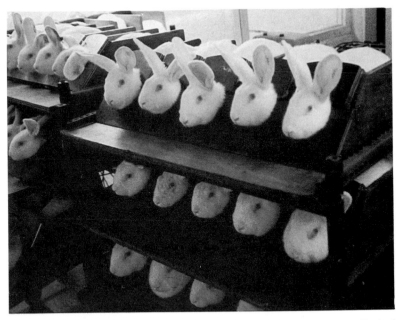

Above *These rabbits are being used for eye irritancy tests at Britain's Toxicol Laboratories.* (© Richard Beggs)

Below *This pig has undergone brain surgery during agricultural experiments at the Institute of Animal Physiology, Babraham, Cambridgeshire.* (© Northern Animal Liberation League)

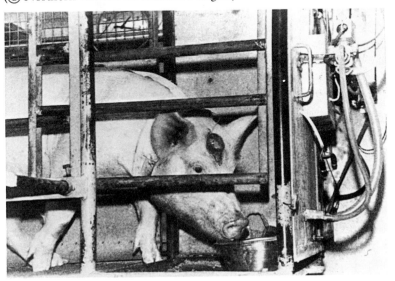

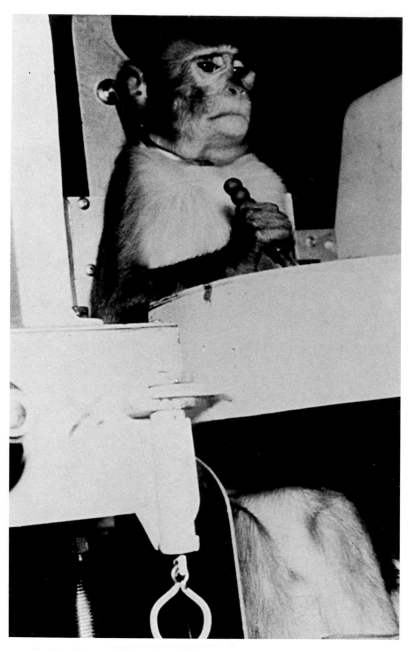

A rhesus monkey operating the Primate Equilibrium Platform during military experiments at the Brookes Air Force Base in Texas, USA.

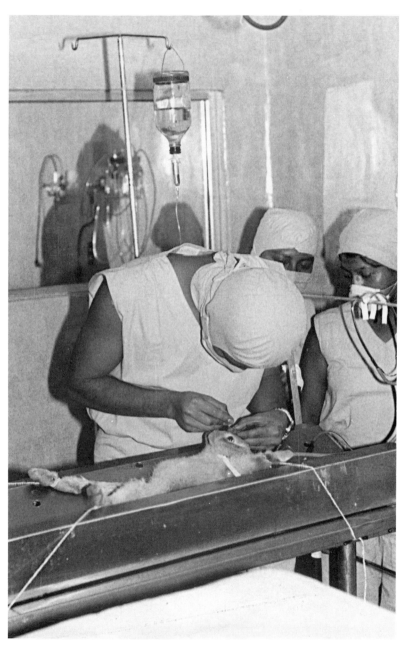

Live rabbit undergoing experimentation in a Spanish laboratory. (© Brian Gunn/IAAPEA)

Above *Life inside a British contract research laboratory. Life Science, Occold, Suffolk.* (© Eastern Daily Press)

Below *Primates seeking social contact in an attempt to overcome the loneliness of their cages. Edward Taubs Laboratories, Maryland, USA.* (Photograph courtesy of PETA, Washington DC, USA)

Above *Britches, an infant monkey subject to sight deprivation experiments at the Riverside Laboratories, University of California, USA.*

Below *Britches after being rescued by the Animal Liberation Front. A survey of the scientific literature between 1983 and 1987, revealed 156 published papers on sight deprivation in the world's laboratories.* (Photographs courtesy of PETA, Washington DC, USA)

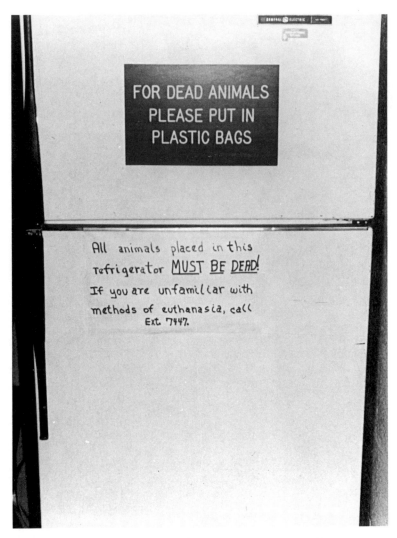

The disposable tools of research. (© Brian Gunn/IAAPEA)

larynx. A laser beam was focussed on each tube until it penetrated, resulting in a blowtorch-type airway fire in the dog's throat. According to the report, '. . . the PVC tube was readily penetrated by the laser and produced an intense flame upon ignition. Black, carbonaceous debris was diffusely deposited in the [trachea] following a fire with the PVC tube. Sections of the mid trachea demonstrated acute inflammation with ulceration . . .' The scientists concluded that there is no laser-safe tube. (R. H. Ossoff, J. A. Duncavage, T. S. Eisenman and M. S. Karlan, 'Comparison of Tracheal Damage from Laser-Ignited Endotracheal Tube Fires', *Annals of Otology, Rhinology and Laryngology*, 333-336, volume 92, 1983. Experiments carried out at Northwestern University Medical School, Chicago and the Medical College of Wisconsin, Milwaukee.)

5 Decapitation of Pig Foetuses

Pig foetuses were beheaded whilst still in their mother's womb to measure the effects on various body chemicals. The foetuses were decapitated at 45 days gestation. At 110 days gestation a total of ten pairs of decapitated and control foetuses were removed from the uterus for measurements to be taken. (R. J. Martin, D. R. Campion, G. J. Hauswan and J. H. Gahagan 'Serum Hormones and Metabolites in Fetally Decapitated Pigs' *Growth*, 158-165, volume 48, 1984. Experiments carried out at University of Georgia and United States Department of Agriculture.)

6 Guinea Pigs Dosed with Soot

Guinea pigs were subjected to toxicity tests with soot derived from a transformer fire. Chemical analysis of the soot, which resulted from a fire in the State Office Building in Binghamton, New York in 1981, had already shown that it contained toxic substances. Nevertheless the animals were dosed with the soot every day for up to 90 days. At the highest dose level administered, 35 per cent of the animals had died by day 32 with all the remaining animals close to death. Adverse effects included loss of weight, emaciation, moderate to severe pneumonia, and liver and kidney damage. (A. P. Decaprio, D. N. McMartin, J. B. Silkworth, R. Rej, R. Pause and L. S. Kaminsky, 'Subchronic Oral Toxicity in Guinea Pigs of Soot from a Polychlorinated Biphenyl-Containing Transformer

Fire', *Toxicology and Applied Pharmacology*, 308-322, volume 68, 1983. Experiments carried out at Centre for Laboratories and Research, New York State Department of Health, Albany.)

7 Baboons poisoned with nerve gas

Baboons were exposed to the nerve gas soman in order to investigate the resulting effects on behaviour and performance. Following exposure, some of the animals experienced convulsions. According to the authors, one of the animals went into '. . . a crouched position, holding his midsection as if in discomfort' following an injection. The scientists concluded that some of the behavioural effects may have been caused by irreparable brain damage. The animals had originally been used in earlier soman experiments to study effects on performance. (E. M. Gause, R. J. Hartmann, B. Z. Leal and I. Geller, 'Neurobehavioural Effects of Repeated Sublethal Soman in Primates', *Pharmacology Biochemistry and Behaviour*, 1003-1012, volume 23, 1985. Experiments carried out at Southwest Foundation for Biomedical Research, San Antonio.)

8 Stomach Ulcers Induced in Rats

Experiments were carried out to investigate how the rat's defensive reactions influence attempts to induce stomach ulcers. In the first experiment rats were restrained in two different sizes of perspex tubes and simultaneously each animal was given 480 3-milliamp electric shocks over a 6-hour period. The animals were then killed and their stomachs examined. Rats restrained in the smaller tube had more ulcers than those able to struggle during their restraint. In the second experiment, restrained and shocked animals who were given the opportunity to gnaw had fewer gastric ulcers. (M. N. Guile and N. B. McCutcheon, 'Prepared Responses and Gastric Lesions in Rats', *Physiological Psychology*, 480-482, volume 8, 1980. Experiments carried out at State University of New York.)

9 Drugs Influence Fighting Fish

Experiments were undertaken to see the effect of the drug pentobarbital sodium on attack behaviour in male Siamese Fighting Fish. The authors note that virtually all the past research on anti-aggressive effects of the drug has focussed on rodents. It was found that high concentrations of the drug reduced fighting whilst not affecting sexual arousal. (M. H. Figler and B. J. Klauenberg, 'Pentobarbital Sodium and Attack Behaviour in Male Fighting Fish', *Psychopharmacology*, 207-208, volume 69, 1980. Experiments carried out at Towson State University, Baltimore.)

10 Induced Bronchitis in Dogs
Bronchitis was artificially induced in mongrel dogs to investigate inflammation and airway obstruction caused by the condition. The animals were exposed to sulphur dioxide for 6-18 months when surviving animals developed chronic coughs and had difficulty breathing. The authors note that their findings '. . . confirm and quantitatively extend the work of others.' (J. Seltzer, P. D. Scanlon, J. M. Drazen, R. H. Ingram and L. Reid, 'Morphologic Correlation of Physiologic Changes Caused by SO_2-induced Bronchitis in Dogs', *American Review of Respiratory Diseases*, 790-797, volume 129, 1984. Experiments carried out at Harvard School of Public Health, Brigham and Women's Hospital, Harvard Medical School and the Children's Hospital Medical Centre, Boston.)

11 Beagles Injected with Plutonium
Experiments were carried out in an attempt to simulate the hand wounds experienced by workers contaminated with plutonium. Eighteen beagle dogs were injected with plutonium in their forepaws and observed for up to eight years. Surviving animals experienced hair loss, thickening of the skin at implant sites, abcesses, fever and tumours. The authors concluded that dogs and people were different in the way the radioactive plutonium spread throughout the body. (G. E. Dagle, R. W. Bristline, J. L. Lebel and R. L. Watters, 'Plutonium-Induced Wounds in Beagles', *Health Physics*, 73-84, volume 47, 1984. Experiments carried out at the US Department of Energy and Colorado State University.)

12 Baby Lambs Subjected to Stress
Experiments were carried out on baby lambs to investigate the impact of deliberately induced stress on their subsequent development. Three times a week for the first five weeks after birth, lambs were removed from the flock and exposed to stress. This involved maternal separation, handling and electric shocks.

The shocks were administered whilst the animals were suspended in a hammock. The authors note that each lamb responded to the experimental procedure with 'Intense vocalization.' When the lambs were five months old they were exposed to more stress to see if their response was in any way dependant on their earlier experience. (G. P. Moberg and V. A. Wood, 'Neonatal Stress in Lambs: Behavioural and Physiological Responses', *Developmental Psychobiology*, 155-162, volume 14, 1981. Experiments carried out at University of California, Davis.)

13 Kittens Deprived of Sight

Experiments were performed to study the effect of sight deprivation on nerve cell changes in the cat's brain. Prior to the time when their eyes would naturally open, 14 kittens had both their eyes stitched closed. Most of the animals were between 6.5 and 10 months old when brain cell measurements were made, after which they were killed. The experimental procedure ensured that the kittens spent their whole lives in darkness. As the scientists pointed out, 'None of the animals used for the study had any lid openings that allowed visualization of the underlying cornea.' (P. D. Spear, L. Tong and C. Sawyer, 'Effects of Binocular Deprivation on Responses of Cells in Cat's Lateral Suprasylvian Visual Cortex', *Journal of Neurophysiology*, 366-382, volume 49, 1983. Experiments carried out at University of Wisconsin.)

14 Monkeys Exposed to Neutron Radiation

Experiments were carried out with rhesus monkeys to study the effects of neutron radiation. In order to have some means of comparing performance before and after exposure, the animals were first trained to avoid electric shocks by pressing a lever. They were then taken to the White Sands Fast Burst Reactor Facility where they were exposed to neutron radiation. Following exposure, some of the animals showed a decrease in correct responses whilst others showed increased reaction times. Two of the animals died prematurely, with the others all developing radiation sickness. The authors suggest further experiments with higher doses of neutron radiation. (M. G. Yochmowitz, G. C. Brown and K. A. Hardy, 'Performance Following a 500-675 Rad Neutron Pulse', *Aviation, Space and*

Environmental Medicine, 525-533, volume 56, 1985. Experiments carried out at Brookes Airforce Base, Texas.)

15 Amputation of Monkeys' Fingers
Eight adult owl monkeys had fingers amputated in experiments designed to see how the brain perceives parts of the body. After amputation care was taken to prevent any regeneration of nerves around the stumps. Following recovery from the surgery, their brains were examined when it was found that the monkeys' perception had changed to take account of the loss of fingers. (M. M. Merzenich, et al, 'Somatosensory Corticol Map Changes Following Digit Amputation in Adult Monkeys', *Journal of Comparative Neurology*, 591-605, volume 224, 1984. Experiments carried out at University of California, San Francisco.)

'Statistics of Experiments on Living Animals in Great Britain 1986' (Command no. 187, HMSO, 1987)

Experiments by type and primary purpose

Number of experiments

Type of experiment	Study of normal or abnormal body structure or function	To select, develop or study the use, etc., of medical, dental and veterinary products and appliances	Development of transplant techniques	To select, develop or study the use, hazards or safety of:										For other purposes	For more than one purpose	Total
				Plant pesticides including fungicides	Herbicides or substances modifying plant growth	Substances used in industry	Substances used in the household	Cosmetics and toiletries	Food additives	Tobacco and its substitutes	Injurious plants or metazoan animals and their toxins	General environmental pollutants	To demonstrate known facts (Certificate C)(a)			
Acute toxicity tests	5,137	179,635	88	23,306	9,004	40,347	1,992	4,643	5,096	–	847	24,547	–	29,449	12,234	336,325
Sub-acute or chronic toxicity tests	3,681	88,209	–	19,319	7,580	6,418	1,808	1,885	1,898	170	695	6,414	–	4,898	3,333	146,308
Teratological tests	2,543	10,946	–	2,852	617	368	–	150	494	–	–	122	–	1,304	36	19,432
Distribution metabolism, excretion and residue tests of substances	23,104	79,770	168	2,046	1,102	1,841	–	–	585	–	39	1,507	26	5,168	5,434	120,790
More than one of the above types of experiment	3,171	14,991	173	–	60	462	–	–	150	–	–	483	3	4,379	21,477	45,349
Total of types specified above	37,636	373,551	429	47,523	18,363	49,436	3,800	6,678	8,223	170	1,581	33,073	29	45,198	42,514	668,204
Other types of experiment	666,718	1,212,524	12,058	7,778	5,202	22,714	5,509	8,974	765	582	228	2,439	999	485,859	11,498	2,443,847
Total	704,354	1,586,075	12,487	55,301	23,565	72,150	9,309	15,652	8,988	752	1,809	35,512	1,028	531,057	54,012	3,112,051

(a) Experiments under Certificate C were performed under anaesthesia and the animal had to be killed before it recovered from the influence of the anaesthetic (author's note)

Experiments by species of animal and concern with neoplasia (cancer) (Great Britain 1986)

Number of experiments

Species of animal	Concern with neoplasia				Total
	Intentional induction of neoplasia	Carcino- genicity screening(a)	Attempted therapy or prophylaxis by any means of natural or induced neoplasms	None	
Mouse	33,072	22,101	63,929	1,503,036	1,622,138
Rat	6,788	22,656	9,776	790,939	830,159
Guinea-pig	89	44	528	131,775	132,436
Other rodent	538	486	429	29,078	30,531
Rabbit	40	56	110	129,173	129,379
Primate	16	7	19	5,593	5,635
Cat	18	–	–	5,233	5,251
Dog	–	–	–	12,901	12,901
Other carnivore	–	–	–	2,635	2,635
Horse, donkey or crossbred	–	–	–	659	659
Other ungulate	30	–	28	32,870	32,928
Other mammal	–	–	–	2,660	2,660
Bird	699	–	422	144,467	145,588
Reptile or amphibian	–	–	–	10,701	10,701
Fish	60	–	–	148,390	148,450
Total	41,350	45,350	75,241	2,950,110	3,112,051

(a) Includes only those experiments in which the screening for possible or known carcinogenic activity was the intention at the outset.

Excludes experiments in which such screening was not the initial intention but where a neoplasm nevertheless occurred.

Experiments by technique used and use of anaesthesia
(Great Britain 1986)

Number of experiments

Technique used		Use of anaesthesia			Total
		No anaesthesia (a)	Anaesthesia for part of experiment	Anaesthesia for whole experiment	
Application of substance to the eye		8,669	2,304	290	11,263
Other interferences with any of the special senses, or the brain centres controlling them:	for behavioural studies	989	449	–	1,438
	for other purposes	311	4,300	1,427	6,038
Interference with the central nervous system (other than centres controlling the special senses):	for behavioural studies	2,854[b]	26,253	–	29,107
	for other purposes	10,133[b]	23,339	12,491	45,963
Use of aversive stimuli, electrical or other:	for behavioural training	35,674	177	–	35,851
	for inducing a state of psychological stress, integral to experiment	3,524	8	–	3,532
Induction by any other means of a state of psychological stress integral to the experiment		22,126	369	–	22,495
Exposure to ionising radiation (other than burning by this means)		65,141	21,010	1,535	87,686
Burning or scalding by any means		452[c]	137	2,213	2,802
Infliction of physical trauma to simulate human injury, other than by burning or scalding		12[d]	3,390	568	3,970
Inhalation		25,834	9,399	3,909	39,142
					(Continued)

Number of experiments

Technique used	Use of anaesthesia			Total
	No anaesthesia (a)	Anaesthesia for part of experiment	Anaesthesia for whole experiment	
More than one of the above techniques	5,208	6,540	1,142	12,890
Total of the above techniques	180,927	97,675	23,575	302,177
Other techniques	2,094,973	485,587	229,314	2,809,874
Total	2,275,900	583,262	252,889	3,112,051

(a) Includes those experiments in which the subject of the study is the anaesthetic agent itself.
(b) In these experiments the interference with the central nervous system was minimal and was of a type for which anaesthesia was inappropriate and could have been harmful.
(c) These experiments were concerned with exposure to ultra-violet rays to simulate sunburn or to induce erythema for other purposes.
(d) These experiments involved exposure to hyperbaric conditions.

Relation of pain to experiment in 1986

Experiments by species of animal and use of anaesthesia (Great Britain 1986)

Number of experiments

Species of animal	Use of anaesthesia			Total
	No anaesthesia (a)	Anaesthesia for part of experiment	Anaesthesia for whole experiment	
Mouse	1,264,444	249,900	107,794	1,622,138
Rat	487,777	237,910	104,472	830,159
Guinea-pig	101,600	15,099	15,737	132,436
Other rodent	15,174	13,000	2,357	30,531
Rabbit	115,444	7,847	6,088	129,379
Primate	3,910	1,241	484	5,635
Cat	817	505	3,929	5,251
Dog	6,812	3,112	2,977	12,901
Other carnivore	503	1,295	837	2,635
Horse, donkey or crossbred	480	138	41	659
Other ungulate	25,350	6,234	1,344	32,928
Other mammal	1,837	480	343	2,660
Bird	140,211	3,948	1,429	145,588
Reptile or amphibian	2,239	5,795	2,667	10,701
Fish	109,302	36,758	2,390	148,450
Total	2,275,900	583,262	252,889	3,112,051

(a) Includes those experiments in which the subject of the study is the anaesthetic agent itself.

Number of licensees experimenting by type of registered place in 1986

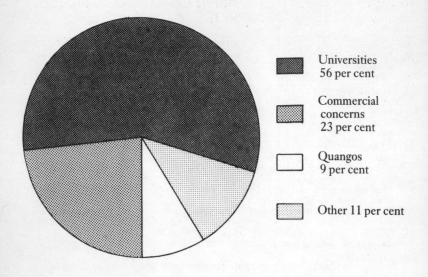

Universities
56 per cent

Commercial concerns
23 per cent

Quangos
9 per cent

Other 11 per cent

Number of experiments by type of registered place in 1986

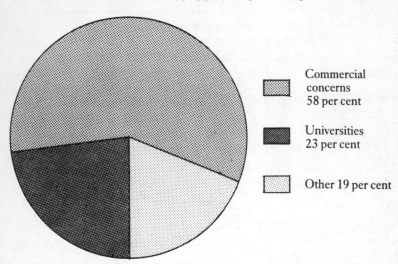

Commercial
concerns
58 per cent

Universities
23 per cent

Other 19 per cent

Further Reading

Peter Singer, *Animal Liberation* (Thorsons, 1983)

Richard D. Ryder, *Victims of Science* (National Anti-Vivisection Society and Centaur Press, 1983)

Tom Regan, *The Case for Animal Rights* (Routledge & Kegan Paul, 1983)

Arabella Melville and Colin Johnson, *Cured to Death* (New English Library, 1983)

Martin Weitz, *Health Shock* (Hamlyn, 1982)

Dianna Melrose, *Bitter Pills* (Oxfam, 1982)

Charles Medawar/Social Audit, *The Wrong Kind of Medicine* (Consumers Association and Hodder & Stoughton, 1984)

Lesley Doyal and Imogen Pennell, *The Political Economy of Health* (Pluto Press, 1979)

Organizations

UK
Animal Aid
7 Castle Street, Tonbridge, Kent TN9 1BH
Telephone: (0732) 364546

British Union for the Abolition of Vivisection (BUAV)
16a Crane Grove, Islington, London N7 8LB
Telephone: (071) 700 4888

National Anti-Vivisection Society Ltd (NAVS)
51 Harley Street, London W1N 1DD
Telephone: (071) 580 4034

Scottish Anti-Vivisection Society (SAVS)
121 West Regent Street, Glasgow, Scotland
Telephone: (041) 221 2300

The Dr Hadwen Trust for Humane Research
Room 2, 6c Brand Street, Hitchin, Herts
Telephone: (0462) 36819

The Lord Dowding Fund for Humane Research
51 Harley Street, London W1N 1DD
Telephone: (071) 580 4034

USA

American Anti-Vivisection Society
Suite 204, Noble Plaza, 801 Old York Road, Jenkintown, PA 19046

Fund for Animals
12548 Ventura Boulevard, Studio City, CA 91604

National Anti-Vivisection Society
100E Ohio Street, Chicago, IL 60611

New England Anti-Vivisection Society
Suite 850, 333 Washington Street, Boston, Mass. 02108

People for the Ethical Treatment of Animals (PETA)
PO Box 42516, Washington DC 20015

American Fund for Alternatives to Animal Research (AFAAR)
175 West 12th Street, New York 10011

Physicians Committee for Responsible Medicine
PO Box 6322
Washington DC 20015

Worldwide

International Association Against Painful Experiments on Animals (IAAPEA)
PO Box 215, St Albans, Herts AL3 4RD, England
Telephone: (IDD) (0727) 35386
The IAAPEA has over 60 member societies operating in over 30 countries and has consultative status with the United Nations Economic & Social Council.

INDEX